NOMENCLATURE *of* THE ORGANIC COMPOUNDS *for* the BEGINNERS

AF633422

NOMENCLATURE *of* THE ORGANIC COMPOUNDS *for* the BEGINNERS

Tanya Das MSc (JU), PhD (Pune University)
Associate Professor, Chemistry Department
D A College, Garia, Kolkata, West Bengal

Amalendu Banerjee MSc (CU), DPhil (CU)
Retired Professor of Chemistry
Jadavpur University, Kolkata, West Bengal
Ex-Hon. Lecturer, RKMR College, Narendrapur
and
Ex-Hon. Teacher, Subhasini Balika Sikshalay
South 24 Pgs, West Bengal

New Central Book Agency (P) Ltd
LONDON
HYDERABAD ERNAKULAM BHUBANESWAR
NEW DELHI KOLKATA PUNE GUWAHATI

NCBA

REGD OFFICE
8/1 Chintamoni Das Lane, Kolkata 700 009, India
email: ncbapvtltd@eth.net

OVERSEAS
NCBA (UK) Ltd, 149 Park Avenue North, Northampton, NN3 2HY, UK
email: ncbauk@yahoo.com

EXPORT
NCBA Exports Pvt. Ltd, 212 Shahpur Jat, New Delhi 110 049
email: ncbaexports@ncbapvtltd.com

BRANCHES

2/27 Ansari Road, 1st Floor
Daryaganj, New Delhi 110 002
email: ncbadel@ncbapvtltd.com

Shop Nos. 3 & 4, Vinayak Towers
681/B Budhwar Peth, Appa Balwant Chowk
Pune 411 002
email: ncbapune@ncbapvtltd.com

Shop No. 15, Bharati Towers, Ground Floor
Forest Park
Bhubaneswar 751 001
email: ncbabub@ncbapvtltd.com

House No. 3-1-315, 1st Floor
Nimboliadda, Kachiguda, Hyderabad 500 027
email: ncbahydb@ncbapvtltd.com

GSS Shopping Complex, 1st Floor
Opposite College Post Office, Convent Road
Ernakulam 682 035
0484-4021054/53

Radhamayee Complex (Second Floor)
Opp. SBI, Panbazar, Jaswanta Road
Guwahati 781 001
email: ncbaguwahati@ncbapvtltd.com

NOMENCLATURE OF THE ORGANIC COMPOUNDS FOR THE BEGINNERS • Tanya Das, Amalendu Banerjee

© Copyright reserved by the Authors

Publication, Distribution, and Promotion Rights reserved by the Publisher

All rights reserved. No part of the text in general, and the figures, diagrams, page layout, and cover design in particular, may be reproduced or transmitted in any form or by any means—electronic, mechanical, photocopying, recording, or by any information storage and retrieval system—without the prior written permission of the Publisher

First Published: September 2013
Reprinted: January 2017

PUBLISHER AND TYPESETTER
New Central Book Agency (P) Ltd
8/1 Chintamoni Das Lane, Kolkata 700 009

PRINTER
New Central Book Agency (P) Ltd
Web-Offset Division, Dhulagarh, Sankrail, Howrah

TECHNICAL EDITOR
Dipan Roy

PROJECT TEAM
Swapan Sarkar, Prodip Baidya and Pradip Biswas

ISBN: 978 81 7381 682 6

Price ₹ 265.00

Dedicated to

Late Prof. Dr Pratul Chandra Rakshit

My Teacher

Amalendu Banerjee

Dedicated to

Shri Soumyabrata Das

My Husband

Tanya Das

CONTENTS

PREFACE

Each book needs an introduction which may or may not be read by anybody, but the seemingly unimportant part is, nevertheless, essential as it explains the purpose of writing such a book and for whom it has been written. The authors thereby take this opportunity to get themselves acquainted with the readers to develop a bond and companionship during the pursuit of the common interest and goal in the study of the materials presented in the book.

Nomenclature is the alphabet of the language of organic chemistry. Without a sound knowledge of nomenclatures—particularly the well known and established methods as put forward by the International Union of Pure and Applied Chemistry (commonly called IUPAC system)—nobody should be expected to be a good organic chemist. This area is vast—the teachers at the primary stage do not get much time to cover even a small portion of it and, at the higher level, the teachers take it for granted that the nomenclature of organic compounds have been thoroughly taught beforehand in the undergraduate level. Thus a great gap is developed and the students experience lots of difficulties to follow the classwork in the graduation and even at the post-graduate level. The ultimate result is that although lots of information and facts have been pumped into the students yet the pleasure and contentment in reading an enchanting book on organic chemistry is lacking because the 'alphabet' of the language had not been learnt well beforehand!!

It appeared to us that a book written in the 'read-it-yourself' style should be helpful to the students—particularly at the elementary stage of learning the subject. Such a book must be written in a simple language, containing lots of varied examples and, may be, some repetition to facilitate smooth and uninterrupted reading. Such a book generally does not require the guidance of the teachers excepting on a few occasions. So the beginners could start reading this important topic even before they get themselves admitted to the next higher classes to build up a strong foundation of organic chemistry.

With a view to furnishing such a book attempts were made by the authors to identify the major difficulties of understanding by the beginners. It appeared that they had problems in 'seeing' the molecule from the trivial (common) nomenclature of the compound, thus involving some sort of uninteresting and dull memory work during the study. But if the IUPAC system of nomenclature system is learnt properly the beginners could envisage the molecule which enables them to readily understand—and even predict—some of the chemical reactions of the compound. Thus, when 'camphor' is mentioned, nothing is understood from this 'common' name, only it can be observed as a colourless solid but no chemical properties can be predicted—not even the molecular (and structural) formula! But, in the IUPAC system, when the name of camphor is given as bicyclo [2, 2, 1] 1,7,7–trimethylheptan–2–one, immediately the molecule will be visualised, the molecular formula would be $C_{10}H_{16}O$ (because) the degree of unsaturation is 3 and a keto group is present as is evident from the IUPAC name, optically active compound, on oxidation it may furnish a dicarboxylic acid etc. could be predicted immediately. Of course there are some limitations in the IUPAC system and some advantages in the trivial system of naming the compound as have been discussed in the text (Sec. 4).

The next major problem is to follow the class lectures delivered by the teachers—specially by the foreign speakers in the seminars—because of the different pronunciations of the names of the compounds and chemical terms. Hence an outline of the pronunciation pattern has been added whenever possible. This will enable the beginners to participate in the verbal discussion and oral presentations with full confidence of proper articulation of the names of the organic compounds. In this book the 'anatomy' of the IUPAC nomenclatures of the compounds has been presented with the mark of fragments of the names during articulation and pronunciation guide in some cases (Sec. 7). It is expected that this would help the students infinitely to remember the whole lengths of nomenclature of organic compounds—particularly when the name is very long. An apposite example will illustrate this point. The name is perhydrocyclopentenophenanthrene—it will be shown in the fragmented form as per.hydro.cyclo.penteno.phenan.threne (for the first time but not afterwards), and the pronunciation guide (*pr.*) is furnished as (per, 'e' in are; 'y' in eye and 'o' in over; cyclo, 'y' in eye and 'o' in over; penteno, first 'e' in end and second 'e' in is, also 'o' in over; phenan, 'e' in is, and 'a' in under; threne, 'e' in is and last 'e' is silent)—once the pronunciation of each fragment is known it will not be repeated in the subsequent appearance of them (Sec. 1).

Another noteworthy problem of the students is proper use of locants (whether necessary or not) and the right choice of the suffix when more than one substituents are the candidates (Sec. 20B). Also, the state of saturation of the main carbon chain (called root name) using 'ane, ene, yne and enyne' has been focused (Sec. 12d iii) and a few 'homemade' mnemonics have been incorporated with the hope of making the study less stressful. Thus, 1-butol, 2,2-dimethylpropane, butan-2-one, but-1-en-yne are all wrong naming on the grounds mentioned above—and the students can identify the mistakes immediately (Sec. 13). Moreover, attention of the students has been drawn to the 'grammar' of nomenclature for the compounds in the IUPAC system (Sec. 5).

From the considerations of aforementioned facts this book is written very carefully from the standpoint of young learners who may wish to read independently during the vacation. It is anticipated that due to several reasons there are some errors which will be rectified thankfully as soon as possible as and when they are pointed out. To give this topic a whole-length character—a totality—the different and important classes of organic compounds, which may not be necessary at the beginning, have been briefly incorporated. A list of textbooks, journals and Latin words frequently used has been included at the beginning, and, at the end, an '**Exercise**' has been added to make the students acquainted with the pertinent questions in this field.

The authors are extremely grateful to their family members—Mrs. Reba Das, Dr. (Mrs.) Santa Banerjee, Mr. Niharendu Banerjee, Mr. Janaranjan Chakraborty, Mr. Soumyabrata Das and his son Rajit—for their constant encouragement and help during the preparation of this manuscript. Thanks are also due to Drs. Haraprasad Samaddar, Gopal C. Banerjee, Mohini Mohan Adak, Arijit Chakraborty, Santanu Banerjee, Anuva Putatunda, Rina Ghosh, Sachchidananda Dutt, Arogya Varan Saha; and the reputed teachers of the departments of chemistry of R. K. Mission Residential College, Narendrapur, and D.A. College, Garia, for their ungrudging help and assistance with valuable advice. Finally, one of the authors (A.B.) expresses his appreciation and sincere gratitude to Dr. Kanti Bhusan Baksi, M.D., Dip. Card (Cal), F.A.C.C. (USA), F.C.C.P. (USA) for his expert medical care and encouragement which enabled the author (A.B.) to complete this book as a joint venture.

We also sincerely thank Mr. Amitabha Sen, Director, New Central Book Agency (P) Ltd, for undertaking the publication. We gratefully acknowledge the superb proof correction, helpful criticism and suggestions by Mr. Dipan Roy (Technical Editor), and Mr. K. Samanta for the improvement of this presentation.

The hard labour of the authors would be highly rewarded by the appreciation of the erudite and interested readers.

June 2013
Kolkata

Tanya Das
Amalendu Banerjee

ACKNOWLEDGEMENTS

The authors thank Dr. Terry Renner, Executive Director, IUPAC, for kindly granting the permission, as quoted below, of using the IUPAC conventions and examples in this book:

> 'You may mention the IUPAC naming conventions and give examples as you have indicated in your note, so long as proper credit is given to IUPAC as the copyright holder of this information.
>
> Best wishes for a successful publication.'
>
> From
> Dr. Terry Renner
> Executive Director
> IUPAC, E-mail: iupacoo@intrex.net
> Date: 31.01.11; Time: 1.28 a.m.

The authors also gratefully acknowledge the help of the books and journals mentioned in '**Reference Books and Journals**'—pg. xv) during the preparation of this book. They should be suitable to the students of the 'Freshman' level. But the advanced students must be prepared to consult all these Reference Books and Journals with special attention to those numbered 18 through 21—to improve their understanding and skill in this area.

[A] Reference Books and Journals

1. Brewster, *A useful Model of Optical Activity*, J. Am. Chem. Soc., 5475, 5483, 5493, 81, 1959
2. Cahn, *An Introduction to Chemical Nomenclature*
3. *An Introduction to the Sequence Rule: A system for the Specification of Absolute Configuration*, J. Chem. Educ, 116, 41, 1964
4. *Chemical Abstract*
5. Eliel, *Recent Advances in Stereochemical Nomenclature*, J. Chem, Educ., 163, 48, 1971
6. Finar, *Organic Chemistry*, Vol. 1 and Vol. 2
7. *Handbook of Chemistry and Physics* (American)
8. Hendrickson, Cram, Hammond, *Organic Chemistry*
9. *Lange's Handbook of Chemistry*
10. Morrison and Boyd, *Organic Chemistry*
11. Pine, *Organic Chemistry*
12. Roberts, Caserio, *Basic Principles of Organic Chemistry*
13. Schill, *Catenanes, Rotaxanes and Knots*, 1971
14. Smith and Crystol, *Organic Chemistry*
15. Wiley Interscience, *Topics in Stereochemistry*, Vol. 5, 1970
16. Bruice, *Organic Chemistry* (Third Edition, 2006)
17. Solomons and Fryhle, *Organic Chemistry* (Eighth Edition, 2007)
18. Sen Gupta and Sen Gupta, *An Introduction to Nomenclature of Organic Molecules* (2005)
19. Sen Gupta, *Basic Stereochemistry of Organic Molecules* (2005)
20. Eliel, Wilen and Mander, *Stereochemistry of Organic Compounds* (2010)
21. IUPAC, *Nomenclature of Organic Chemistry*, Sections A, B, C, D, E, F and H (1979), Pergamon Press; *A Guide to IUPAC Nomenclature of Organic Compounds* (1993), Jain-Interscience Press
22. Pal – *Nomenclature of Organic Compounds*

[B] Latin Words (and abbreviations)

1. circa = approximately, about (ca)
2. ditto = as shown just above (do)
3. exempli gratia = for example (e.g.)
4. et alii = and others (et al)
5. et sequens = and what follows (et seq.)
6. et cetera = and other similar things (etc.)
7. id est = that is to say (i.e.)
8. ibid = in the same source (ibidem)
9. idem = the same (id) [used in citations to indicate an author or word that has just been mentioned]
10. idem quod = in the same place of unknown origin (iq)
11. in situ = in the original or appropriate position
12. loco citato = in the place cited (loc. cit)

13. opera citato = in the work already cited (op. cit.)
14. quod vide = see which one (q. v.) [used to direct a reader to another part of a text for further information]
15. ut supra = as above (u.s.)
16. vide infra = see below (v.i.)
17. vide supra = see above (v.s.)
18. videlicet = namely (viz.)
19. vice versa = both ways [with the main items in the preceding statement the other way round]

1. Introduction

Nomenclature is the gateway of organic chemistry. For the organic chemists reading, understanding and verbal communication will be very difficult without the proper knowledge of the nomenclature which supplies many information regarding the chemical properties of the organic compound under consideration.

A very important part is the proper pronunciation or articulation of the names of the compounds during transfer of knowledge verbally from one person to another. Thus, if periodic (*pr.* par.io.dic) acid is uttered as piriodic acid, ester (e as in end; *pr.*es.ter) is pronounced as easter, ister or aster, lead (*pr.* led) as lid and chiral (*pr.* kiral, 'i' as in eye) as siral, syral, cheeral or chairal (ch as in chair), then the audience will be in real trouble in understanding the speaker. There is no need to cite other examples of mispronunciation of words or terms associated with organic chemistry as the necessity of proper pronunciation has been well understood by the readers from the above examples. It should be pointed out here that the pronunciation varies from country to country, as the Australian, English and American pronunciations vary in some cases (not to speak of Japanese, Indian, Chinese-English, etc.), but the overall pattern remains identical, thus enantiomer (*e* in end and *t* in tin), enansiomer, inantiomer ('i' in is) or inansiomer conveys the same meaning but the spelling must be 'enantiomer'. The main difference between American and British pronunciation is generally found in vowels like 'e' and 'i', and '–yl' syllable. Thus, (pronunciation = *pr.*, cf = as in).

	Vowels/ Syllable e/i, yl	Americans (pr.)	British (pr.)	Example	American (pr.)	British (pr.)
[i]	–yl	(il, cf it)	(ail, cf. isle)	Methyl	(methil)	(methail)
[ii]	E/e	(cf. end)	(cf. is)	Ethyl	(ethil)	(ithail)
[iii]	–li	(cf. lie)	(cf. lick)	Alkali	(alkalie)	(alkali)
[iv]	–i	(cf.eye)	(cf. is)	Pipette	(paipet)	(pipett)
[v]	Phe	(cf Fen)	(cf Fig)	Phenyl	(fenil)	(finail)
[vi]	Aryl	(il)	(ail)	Aryl	(aril)	(arail)
[vii]	Vinyl	(il)	(ail)	Vinyl	(vainil)	(vainail)
[viii]	Allyl	(il)	(ail)	Allyl	(alil)	(alail)
[ix]	Nitrile	(il)	(ail)	Nitrile	(naitril)	(naitrail)
[x]	Amide	(id)	(aid)	Amide	(amid)	(amaid)

There are also some differences in spelling of several words between the American school and British school. Only a few examples are given below to illustrate the point.

American School	British School
[i] Sulfur	[i] Sulphur
[ii] Sulfuric Acid	[ii] Sulphuric Acid
[iii] Crystallize	[iii] Crystallise
[iv] Color	[iv] Colour
[v] Aluminum	[v] Aluminium
etc.	etc.

It is needless to say that these differences in spelling and pronunciation are extremely insignificant but it is desirable that the beginners should know it and never mix one type with the other during speaking and writing of scientific articles.

2. Representations of Organic Molecules

It is highly desirable at this stage to be acquainted fully with the different methods of drawing the molecules. These presentations are following drawings (taking, C_5H_{12}, *n*-pentane as example).

[A] $C_5H_{12} = \overset{1}{C}H_3 - \overset{2}{C}H_2 - \overset{3}{C}H_2 - \overset{4}{C}H_2 - \overset{5}{C}H_3$ = Let us call 'long-hand formula'.

[B] do. = 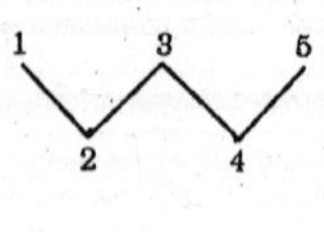 = Let us call 'short-hand formula', this is also known as 'topological' formula

[C] do. = $\overset{1}{C} - \overset{2}{C} - \overset{3}{C} - \overset{4}{C} - \overset{5}{C}$ = This is 'skeletal formula'.

[D] do. = $CH_3(CH_2)_3 - \overset{5}{C}H_3$ = This is 'condensed formula'.

[E] do. = 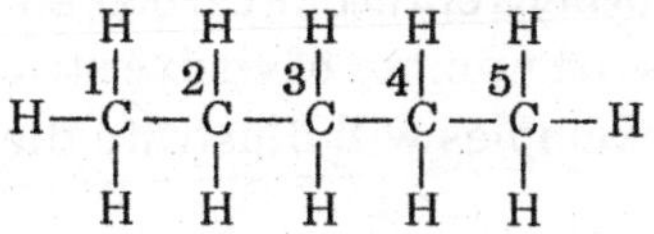 = This is 'structural formula'.

[F] do. = 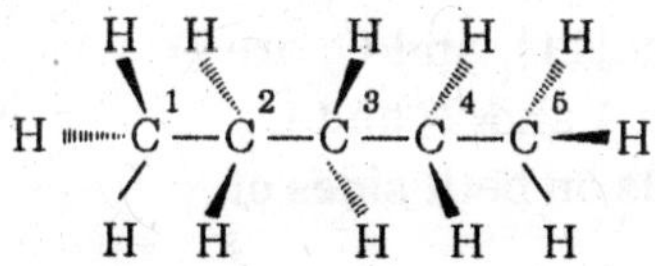 = This is three dimensional 'stereo-structural formula' [▬ β-bond; ┅ α-bond; —on plane bond]

[G] do. = 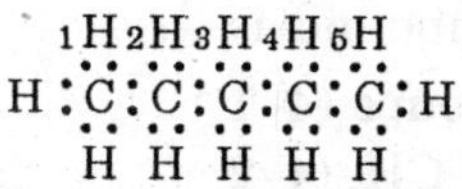 = This is 'Lewis-dot formula' [The octet has been completed]

It is desirable to discuss briefly each one of the above representations.

[A] In the 'long-hand formula' the direct linkage between carbon and hydrogen and other heteroatom (if any) is clearly demonstrated. Moreover, the location of double bond and /or triple bond in the chain is easily identifiable. To some extent the isomers are also understood easily. But this formula writing takes a longer space and time associated with the risk of committing mistakes in the numbers of the bonded atoms as well as the monotony and cumbrous pictures-drawing of larger molecules.

Examples

[a] Isomers of C_5H_{12} in long hand picture

[i] $H_3C—CH_2—CH_2—CH_2—CH_3$ [ii] 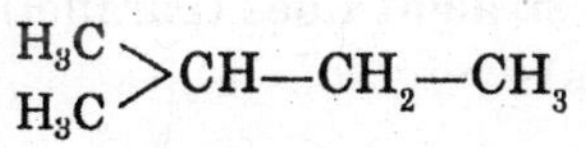 [iii] 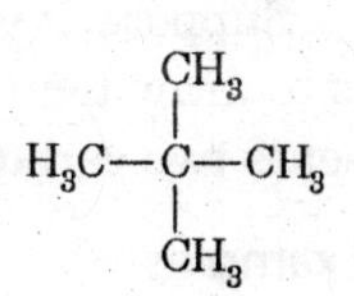

n-pentane (*n* = normal) *iso*-pentane *neo*-pentane

[b] Isomers of C_4H_6 in long-hand picture

[i] $\overset{1}{H_3C}—\overset{2}{C}≡\overset{3}{C}—\overset{4}{CH_3}$ 2-butyne (the triple bond is between C_2 and C_3)

[ii] $\overset{1}{HC}≡\overset{2}{C}—\overset{3}{CH_2}—\overset{4}{CH_3}$ 1-butyne (the triple bond is between C_1 and C_2)

[iii] $\overset{1}{H_2C}=\overset{2}{CH}—\overset{3}{CH}=\overset{4}{CH_2}$ 1,3-butadiene (two double bonds, 1st between C_1 and C_2 and the 2nd between C_3 and C_4)

and many more structures

[B] In the 'short-hand formula' the carbons and hydrogens are not shown in their respective symbols, but the heteroatoms. (O, N, S, P, Halogenes, etc.) are shown and the hydrogens attached directly to them are slso shown symbolically. The carbon chain is drawn in

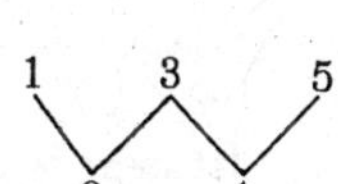

n-pentane

a zig-zag way, the terminations and each corner represent carbon and the number of hydrogen is calculated from (4–x, where x is the number of visible straight lines representing the valency bonds). The following examples will illustrate the point.

[i] Isomers of C_5H_{12} in short-hand formula.

[a] **Explanation**: Five carbon atoms, marked 1 through 5 are connected in a zig-zag way. Terminals C_1 and C_5 are CH_3 groups (4–1, visible bonds between 1 and 2) and C_2, C_3, C_4 are CH_2 groups (4–2, visible bonds on both sides of the corners).

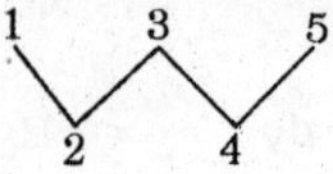

[a]

n-pentane

[b] **Explanation**: *n*-chain does not have any branching and its isomers have one or more branching, i.e., C and H containing groups as substituent(s). When the termination is $(CH_3)_2CH–$, i.e., 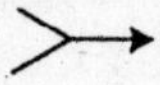 to carbon, the compound is termed as *iso*. Here, C_1, C_4 and C_5 are CH_3 groups (4–1 bonds), while C_2 and C_3 are CH (4-3-bonds) and CH_2 (4–2 bonds), respectively.

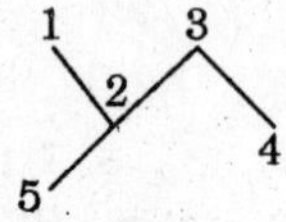

[b]

iso-pentane

[c] **Explanation**: Here C_1, C_3, C_4 and C_5 are CH_3 groups (4–1 bond) while the central C_2 does not contain H (This is characteristics of a *neo*-carbon). So it is concluded that in order to get a *neo*-compound, there must be at least (i.e., minimum number) five carbon atoms in the molecule. The C_2 is *neo* carbon which is also called the quaternary carbon.

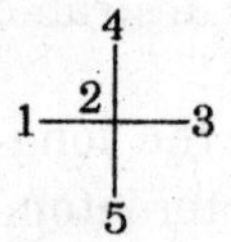

[c]

neo-pentane

The short-hand formula has the advantages like [i] time saving and space saving, [ii] less chance of committing mistakes, [iii] less crowded, so easy to understand and easy to remember. In the following is given an easy way of drawing short-hand structures.

Suppose, it is required the short-hand structure of $C_{20}H_{42}$ to be drawn. The quick method is to draw ten (20 ÷ 2) straight lines (parallel) as shown below (each straight line represents two carbon atoms).

Examples

[i]

1st step (20 carbons) 2nd step (eicosane, $C_{20}H_{42}$)

In even numbered carbon chain the terminals will be on opposite direction while in odd numbered carbon chain the terminals will be on the same direction. The last carbon in the example [ii] is positioned as a dot.

[ii] To draw $C_{11}H_{24}$ in short-hand

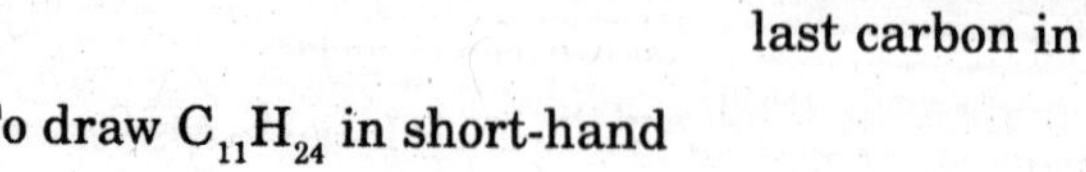

1st step (5 lines + 1 dot) 2nd step ($C_{11}H_{24}$ undecane, terminals on the same direction)

It is needles to say that the 1st and 2nd steps can be combined together (in the same drawing) one after another thus eliminating the separate drawings.

The heteroatoms are shown in their respective symbols with the symbol of hydrogen connected to the heteroatom. The following examples will make the above statements well understood. (iii to xii). [L.H. → S.H.]

	Long-hand (L.H.)		Short-hand (S.H.)
[iii]	$CH_3—CH_2—OH$ (C$_2$H$_6$O, ethanol)	⟶	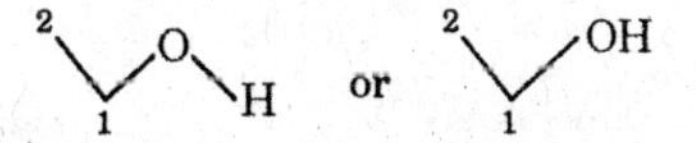
[iv]	$CH_3—CH(OH)—CH_3$ (C$_3$H$_8$O, *iso*-propyl alchohol)	⟶	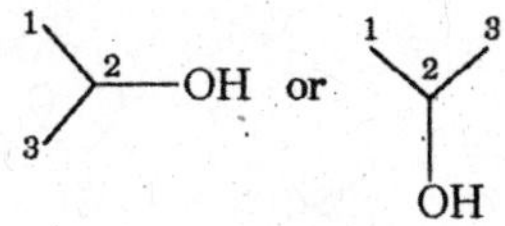
[v]	$H_3C—C(CH_3)_2—CH_2—NH_2$	⟶	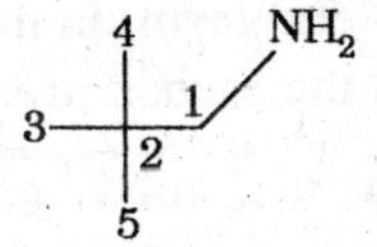
[vi]	$CH_3—CH_2—CH_2—O—CH_2—CH_3$	⟶	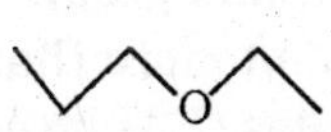
[vii]	$CH_3—CH_2—CH_2—N(CH_3)_2$	⟶	
[viii]	$CH_3—CH_2—CH_2—N(CH_3)H$	⟶	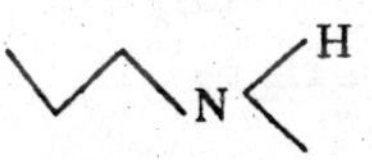
[ix]	$CH_3—C{\equiv}C—CH_3$	⟶	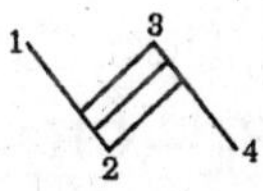 i.e. 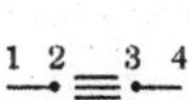
[x]	$CH{\equiv}C—CH_2—CH_3$	⟶	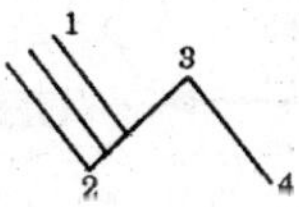 i.e. 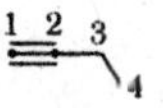
[xi]	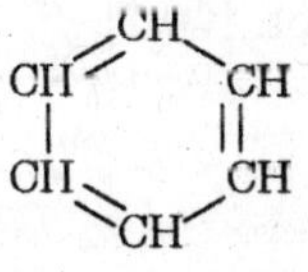	⟶	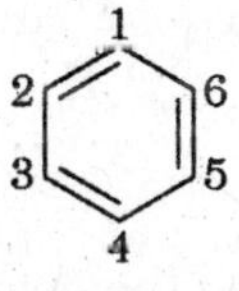
[xii]	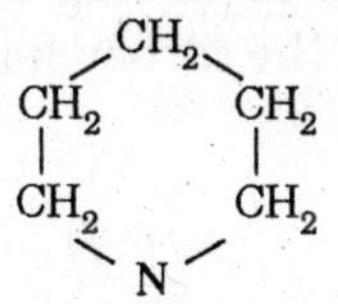	⟶	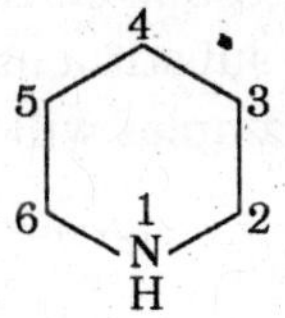

[C] **Skeletal formula**: In this method only the carbons are shown in the chain and in the whole molecule. Thus, the isomers of C_5H_{12} are shown below. It should be noted that the skeletal formula can be represented in several ways for the same compound (e.g. ii).

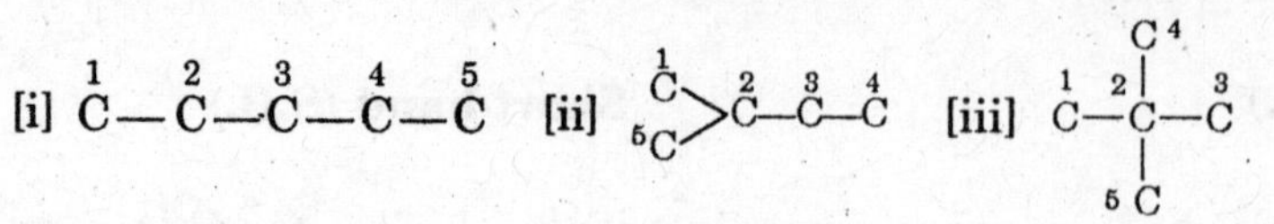

[C_2—is *neo* (or quaternary)]

(*n*-pentane) (*iso*-pentane) (*neo*-pentane)

[ii] Various representations of *iso*-pentane in skeletal formula (all forms are same).

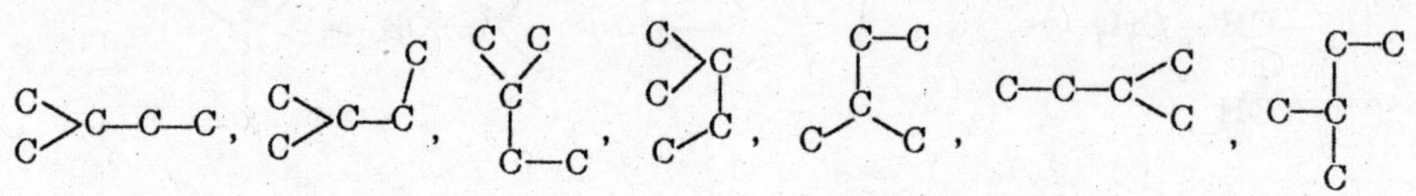

The drawing of skeletal formula may prove itself suitable to show a particular attack of reagent on a specific carbon atom.

[D] **Condenesd formula**: Sometimes several CH_2 groups occur one after another in the chain and /or a functional group (Q) in the molecule which is shown as $(CH_2)_n$ and Q_n (n = whole number and more than 1) in the condensed form. But in some molecules the condensed formulas (L.H. to condensed form) pose some problems for easy understanding.

[i] $\overset{1}{C}H_3-\overset{2}{C}H_2-\overset{3}{C}H_2-\overset{4}{C}H_2-\overset{5}{C}H_2-\overset{6}{C}H_3 \rightarrow \overset{1}{C}H_3-(CH_2)_4-\overset{6}{C}H_3$

(*n*-hexane) (easily understood)

[ii] $HOO\overset{1}{C}-\overset{2}{C}H_2-\overset{3}{C}H_2-\overset{4}{C}H_2-\overset{5}{C}H_2-\overset{6}{C}O_2H \longrightarrow HO_2C-(CH_2)_4-CO_2H$ (do.)

(adipic acid) (Q = CO_2H group)

$\longrightarrow (CH_2)_4(CO_2H)_2$ (difficult to understand) [Q=2].

[iii] $HOOC-COOH \longrightarrow \begin{matrix} CO_2H \\ | \\ CO_2H \end{matrix}$ (easily understood), but $(CO_2H)_2$ not. [Q=2]

(Oxalic acid)

Generally during drawing the formula in long-hand or condensed form the substituents or the functional groups on the first left-hand carbon is fixed on the left of this carbon, after this these substituents are placed on the right side of the carbon holding them. The following examples will clarify the point.

Proper representation	**Improper representation**
[i] $ClCH_2-COOH$ (Chloroacetic acid)	$CH_2Cl-COOH$ or CH_2Cl-CO_2H
[ii] HO_2C-CO_2H (Oxalic acid)	CO_2H-CO_2H or CO_2H-HO_2C
[iii] $Br-\overset{1}{C}H_2-\overset{2}{C}H_2-Br$ (1,2-dibromoethane)	CH_2Br-CH_2Br or $CH_2-Br-Br-CH_2$ or $Br-CH_2-BrCH_2$

[iv] $\overset{1}{C}H_3-\overset{2}{C}HCl-\overset{3}{C}H(OH)-\overset{4}{C}H_3$ (Cl at C_2 and (OH) at C_3 are on the right side of C, holding them.)

[v] $H_3\overset{1}{C}-\overset{2}{C}H_2-\overset{3}{C}Br(NO_2)-CH_2-CH_3$
[Both Br and (NO_2) groups are on the right side of C_3, holding them. The functional groups are in parenthesis.]

[vi] $\overset{4}{C}H_3-\overset{3}{C}H_2-\overset{2}{C}HCl-\overset{1}{C}H_2OH \equiv HO-\overset{1}{C}H_2-\overset{2}{C}HCl-\overset{3}{C}HCl-\overset{3}{C}H_2-\overset{4}{C}H_3$

[vii] $(H_3C)_2N-\overset{3}{C}H_2-\overset{2}{C}HBr-\overset{1}{C}O_2H \equiv HO_2\overset{1}{C}-\overset{2}{C}HBr-\overset{3}{C}H_2-N(CH_3)_2$

$\equiv HOO\overset{1}{C}-\overset{2}{C}HBr-\overset{3}{C}H_2-NMe_2$

(Me$=CH_3$ Group)

But 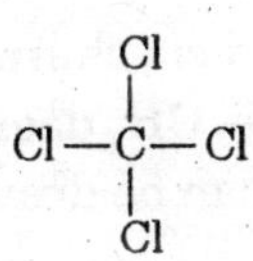 or 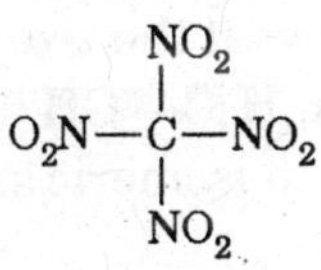 should be CCl_4 or $C(NO_2)_4$,

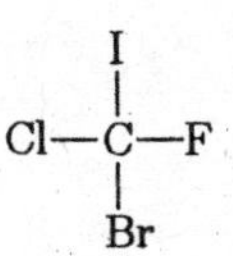 written as CFClBrI.

The heteroatom directly attached to the carbon should be shown clearly. Thus, in 1,2-dinitroethane, since the N is connected directly with the carbon (not the O connected to C) the proper drawing will be

[viii] $O_2N-CH_2-CH_2-NO_2$ and not $NO_2CH_2-CH_2O_2N$ or $NO_2-CH_2-CH_2-NO_2$.

More examples (ix through xiii)

[ix] $HO-\overset{1}{C}H_2-\overset{2}{C}ClBr-\overset{3}{C}H_2-SH$

[x] $HO_3S-\overset{6}{C}H_2-(CH_2)_4-\overset{1}{C}O_2H \equiv HO_2\overset{1}{C}-CH_2-(CH_2)_3-\overset{6}{C}H_2-SO_3H$

[xi] $C_6H_5—CH_2—\overset{1}{C}H_2—NH_2 \equiv H_2N—\overset{1}{C}H_2—CH_2—C_6H_5$

(C_6H_5 is not depicted as H_5C_6)

[xii] $NC—CH_2—CH(CO_2H)—CH_2—CN$, (CN=Cyano),

this is $N\equiv C—CH_2—\overset{\overset{CO_2H}{|}}{C}H—CH_2—C\equiv N$

[xiii] $CN—\overset{4}{C}H_2—\overset{3}{C}H(CO_2H)—\overset{2}{C}H_2—\overset{1}{C}N$ (CN on C_4 is isocyanide);

this is $C{=}N—CH_2—CH(CO_2H)—CH_2—C\equiv N$

[E] **Structural formula**: In this method each atom of the molecule is shown separately bonded with carbon in a two dimensional drawing. Since the exact location of the reacting centre is clearly demonstrated the whole sequence of reaction leading to the product is more readily understood. Thus,

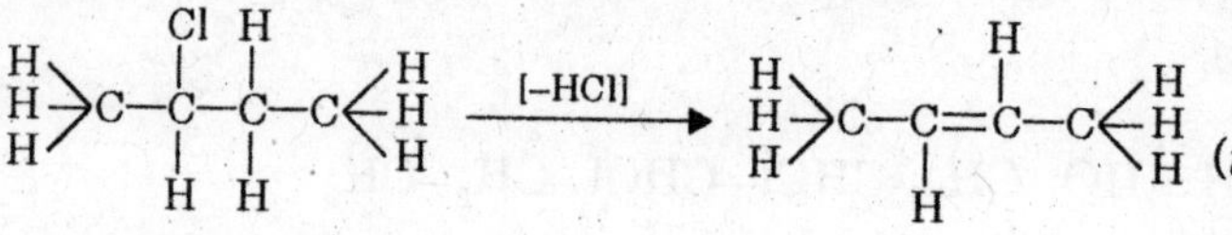

(*trans*–compound is formed)

[i]

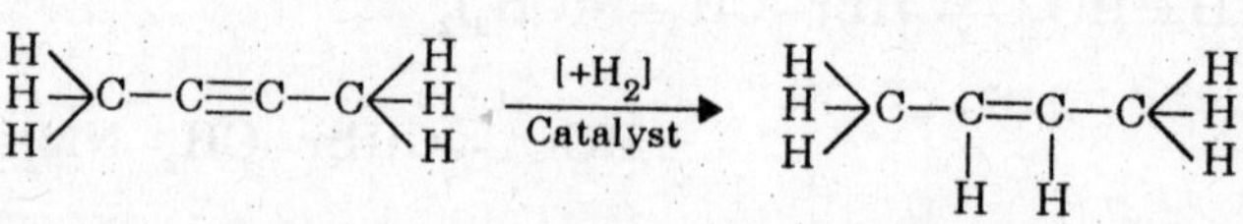

(*cis*–compound is formed)

[ii]

The structural formulae (i) and (ii) immediately reveal the *trans* and *cis*-characters of the two isomers, C_4H_8, but the long-hand formula, $H_3C—CH{=}CH—CH_3$ does not do so. Of course the short-hand formula of this olefin in isomeric state can be drawn with equal ease and advantage. Thus,

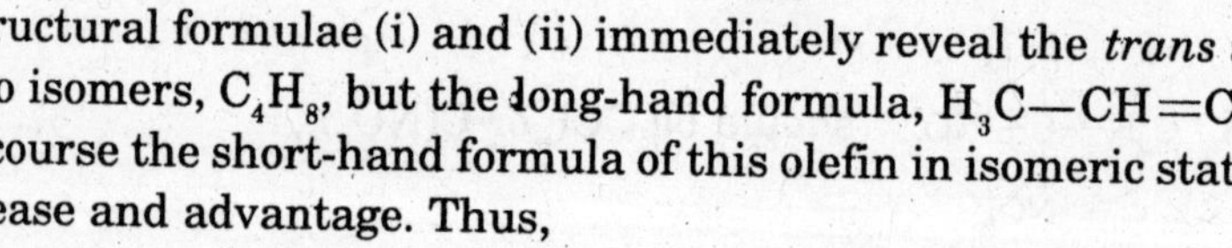

[i] *trans*-butene (homomer) [ii] *cis*-butene (homomer)

[F] **Stereostructural formula**: This is the improved form of structural formula because it is a three-dimensional drawing. So the distance between two atoms can be determined and the physical and chemical properties can be predicted to some extent. This utilises three types of bond representations.

[i] ——, this is *on plane* bond, resting on the plane of the paper.

[ii] ◄, this is β-*bond*, proceeds towards the observer (β=beta).

[iii] IIIII or --------, this is α-bond, proceeds away from the observer (α=alpha).

If the observer changes its position in the opposite direction, the β-bond will change itself into α-bond and vice versa. Generally two bonds of C in sp^3 hybridisation are on-plane, one bond is β and the other is α-bond.

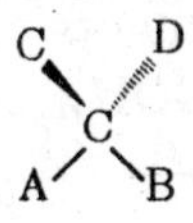

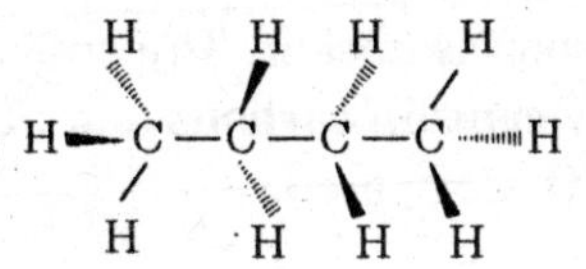

$= C_4H_{10} =$ *n*-butane

[G] **Lewis-dot formula**: Shows the electronic arrangement and environment of each atom of the molecule. Thus,

[i] 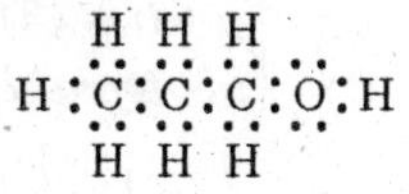

$(CH_3–CH_2–CH_2OH)$

[ii] 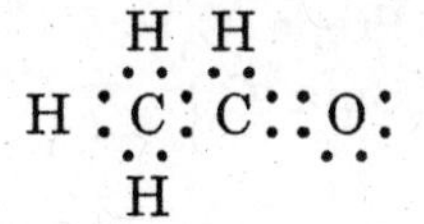

$(CH_3–CHO)$

[iii] 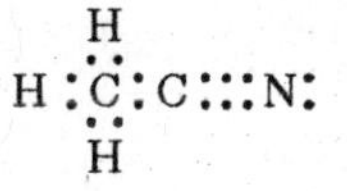

$(CH_3–CN)$

[iv] 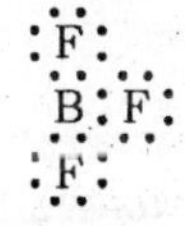

(BF_3)

3. Root Names and Number of Substitutions

The nomenclature of a compound is the exposure of its identity and the linkages of the atoms in the molecule and to some extent expression of the related chemical and physical properties. Better nomenclature gives better exposure of the nature and structure of the molecule.

Selection of the root name of the molecule is a very important step. The root name indicates the largest number of carbon atoms in the carbon chain of the molecule. The first four root names are (i) C_1 = metha (*pr.* 'e' in end and 'a' in after), (ii) C_2 = etha, (iii) C_3 = propa ('o' in old), and (iv) C_4 = buta ('u' in unit).

After C_4 (buta) the root name is expressed by numerals derived from Latin which is applicable to both carbon as well as functional group or substitution on the chain. Thus, penta (C_5), hexa (C_6), hepta (C_7), octa ('o' in on, C_8), nona (C_9), deca (C_{10}), undeca ('u' in ugh, C_{11}), dodeca ('o' in only, C_{12}), trideca ($3+10=C_{13}$), eicosa (C_{20}), heneicosa ('e' in India, 'ei' in I, C_{21}), docosa ($2+20=C_{22}$), triaconta ($3\times10=C_{30}$), pentaconta ($5\times10=C_{50}$), hentetraconta ($1+4\times10=C_{41}$), hecta (C_{100}), kilia (C_{1000}; lia$=10^3$), etc.

The first four numerals *other than carbon* are expressed as mono (1), di (2), tri (3) and tetra (4). These are applicable to denote the total number of substituents [element(s) and / or functional group(s)]. Sometimes di is replaced by bi (2). The substituents (they replace the hydrogens on the carbons of the main chain) are attached directly with these numerals (indicating 'how many' of that substituent) which is called 'multiplying prefix' and the substituent is generally called 'prefix' (pre=before, fix=attach) as they are followed by the root name. The following points should be clearly understood.

[a] Carbon is *present* in the main chain. The longest chain will deliver the 'root name'.

Carbon is *absent*. The substituents may contain carbons (e.g., COOH, CHO, CH_3, etc.).

	Simple functional groups	*Complex functional groups*
[i] Metha (C_1)	[i] mono (1)	
[ii] Etha (C_2)	[ii] di (or bi, 2)	*bis* (2)
[iii] Propa (C_3)	[iii] tri (3)	*tris* (3)
[iv] Buta (C_4)	[iv] tetra (4)	*tetrakis* (4)

After this; no distinction; same multiplying frefix.

↓

[v] Penta (C_5 as well as 5) ... pentakis (5)
[vi] Hexa (C_6 as well as 6)... hexakis (6)

[b] The multiplying prefix is to be used for only *same substitutents*. Thus, five chlorine atoms will be termed as pentachloro, but two bromine atoms and three chlorine atoms should be dibromo-trichloro, three carboxy groups (COOH) will be tricarboxy, four nitro groups ($-NO_2$) will be tetranitro, etc. But when the substituents are complex (the previous examples were all simple) groups like dimethylamino [$-N(CH_3)_2$], then the multiplying prefixes will be *bis* (2), *tris* (3), etc. Thus [$-N(CH_3)_2]_2$ will be *bis* (dimethyl amino), [$-C_6H_4Cl(p)]_3$ will be *tris* (*p*-chlorophenyl), etc. As before the substituents should be identical.

[c] The 'class' of compounds depends on the functional groups which control the chemical and physical properties of the compound. This 'class' is denoted by 'suffix' (attached after the 'root name'). The root name and the suffix are the minimum requirements to name a molecule in IUPAC system.

[d] **Explanations**: The saturated hydrocarbons are denoted by the suffix 'ane' (they do not contain C═C or C≡C bonding). But the unsaturated hydrocarbons are denoted by 'ene' (contains C═C, one or more) and 'yne' (contains C≡C one or, more). The multiplying prefixes are used to express the number of 'ene' or 'yne'. The same molecule may contain both ene and yne as 'enyne' (suffix) but the 'ane' is never connected to either ene or yne or enyne. The general formula of saturated hydrocarbon is C_nH_{2n+2}.

[i] CH_4 is saturated, so suffix is ane, it contains only 1 carbon, so metha. The name is methane (not 'mono ane' or 'monane'), 'Metha' (root name) and 'ane' (suffix, character of compound) can make a complete name of the compound.

[ii] C_2H_6 is saturated, so suffix is 'ane', it contains only 2 carbons, so etha. The name of the compound is ethane (not diane or biane). Etha (root name) and 'ane' (suffix, character of the compound is saturate) make the complete name.

[iii] C_3H_8 (n=3 in general formula), is saturated (so suffix is 'ane'), it contains three carbons, hence 'propa' (root name). The complete name is propane (not triane).

[iv] C_4H_{10} (n=4), saturated (suffix 'ane') attached after buta (C=4), making the complete name butane. Since ane, ene and yne when connected to a root name can give a complete name, these suffixes will be termed here as primary or major suffix. Butane will not be named as tetrane.

[v] $C_3H_5Cl_3$ will be trichloropropane (needs 'locant', vide infra) not trichlorotriane, as well as not propachloropropane. $C_4H_8Cl_2$ will be dichlorobutane (not ethachlorobutane or dichlorotetrane).

[vi] After buta there is no difference between the carbon content of the chain and the number of substituents [vide 3, (a)]. Thus $C_5H_6Br_6$ will be hexabromopentane and $C_5H_7Cl_5$ (pentachloropentane), $C_6H_{12}(NO_2)_2$ (dinitrohexane), $C_6H_8I_6$ (hexaiodohexane), etc.

4. Background of Nomenclature of Organic Compounds

From the very beginning there was the need for naming the organic compounds. Initially the compounds were named from the source, e.g., formic acid, acetic acid, citric acid, benzene, butane, methyl alcohol, etc. Such system furnished the trivial names or common names. But it was found to be inadequate very soon and International Chemical Congress started to develop a systematic method of naming the Organic Compounds (1889). Their suggestions were submitted at Geneva which was accepted as the improved method (1892) but slowly it was realised that there were some defects in this Geneva nomenclature. So a new organisation named International Union of Chemistry was established (1919) to improve further the nomenclature system. They published the IUC system (1930) but later established another organisation named International Union of Pure and Applied Chemistry (abbreviated as IUPAC) in 1974. Their method of nomenclature of organic compounds has been accepted universally and enthusiastically.

In spite of the continuous effort of IUPAC to furnish unambiguous names of organic compounds, some difficulty still exists. Sometimes the IUPAC nomenclature is very complex and difficult to understand. As for example, the name of $C_{10}H_8$ as naphthalene is very common and well understood but its IUPAC name bicyclo [4, 4, 0] 1(6), 2, 4, 7,9-decapentaene is not so easily understandable. So in this case the trivial name is more suitable. Therefore, the beginners should know both trivial and IUPAC system of nomenclature with special attention to the latter system. When the structure of the compound is too complex and nomenclature is difficult, the compound is marked by Roman numerals (I, II, III, etc) or Arabic numerals (1, 2, 3, etc) to make the presentation good. The Geneva system of nomenclature may be used to describe a series of reactions starting from trivial name of the substrate (i.e., the starting organic compound, called starting material) to which is added stepwise the (i) name of the functional group with location (i.e., locant), (ii) name of other substituents (if any) with their respective locants, (iii) the changed name of the product related to the change of structure of the substituents. The following example is illustrative.

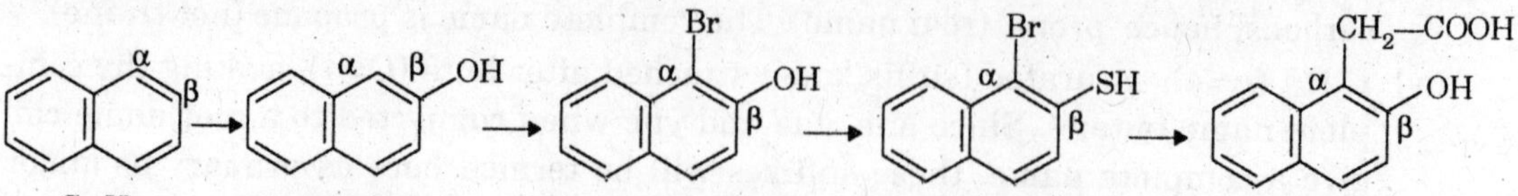

$C_{10}H_8$ (Naphthalene) common name → β-naphthol (locant-β, functional group —OH) → α-bromo-β-naphthol (bromine at α and OH at β of naphthalene) → α-bromo-β-naphthiol (Br at α and SH at β) → β-hydroxynaphthyl -α-acetic acid

5. Grammar of the Nomenclature of Organic Compounds

During the nomenclature of organic compounds by IUPAC method the spelling of the name follows a general rule most of the time. It has been observed before that CH_4 is named methane (C_1 delivers the 'metha' and saturation delivers 'ane' as the suffix) instead of the combined moiety methaane as expected. The grammar dictates when two vowels (or two letters with the pronunciation like those of vowels) appear side by side, the first vowel is to be dropped in most of the cases. Thus, metha+ane=methaane, becomes 'methane' by dropping the first 'a' of the two consecutive 'a's. Similarly propaane, hexaane and octaane become propane, hexane and octane, respectively.

In case of compounds holding C=C framework the suffix is 'ene'. Hence C_2H_4 (general formula is C_nH_{2n}) should be etha (for C_2)+ene (for C=C)=ethaene. The first vowel 'a' from 'ae' is dropped and the name becomes ethene (*pr.* like 'ithin', the last e is silent). Thus, butene (but+ene), hexene (hex+ene), etc. are spelt in this way and the pronunciation (*pr.*) is also changed accordingly.

The compounds holding C≡C get the suffix 'yne' (*pr.* 'y' as in 'idea', a letter sounding like a vowel) Hence C_4H_6 (general formula for yne is C_nH_{2n-2}) is named as butyne (but+yne). Similarly, pentyne, nonyne, decyne, etc. are spelt.

The 'ane', 'ene' and 'yne' are singular. When the last 'e' of 'ane' is present, the 'ane' is pronounced as N (cf. *end*), but when the last 'e' is dropped, as required by the grammar, the pronunciation is changed to 'an' (as in '*un*der'). This pronunciation will guide the spelling and vice-versa.

It is very difficult for the beginners to pronounce the big (sometimes very big) names of the organic compounds. Initially the syllables are denoted in this book by a dot (.) to break the names of the molecules. The speaker will articulate the names stopping for a break at each dot. Once the speaker becomes accustomed to the pronunciation properly by identifying each syllable there would be no need for the dot sign any more.

Examples

[i] The name is cyclopentanoperhydrophenanthrene. Initially for the beginners the name is dotted as cyclo. pentano. perhydro. phenan. threne.

[ii] Octahydroanthracene is given the pronunciation guide here as octa., hydro., anthra., cene., So the naming of cyclooctatetraene poses no problem of pronunciation. Generally the accent is on the first and or second syllable during pronunciation.

More examples of pronunciation guide

[i] Alkyl = al.kyl (stress on al; 'a' in and; kyl is soft)

[ii] Pentachloronitroethane = Penta.chloro.nitro.ethane.

[iii] Hydrocarbon = hydro.carbon.

[iv] Hydroxyl = hy.droxyl.

[v] Olefins = Olefins also (*pr.* Olefinz; cf. dogs = dogz)

[vi] Iodo = i.odo (sometimes pronounced as yodo)

[vii] Ion = i.on (some times sounds like ayon in the middle of a sentence)

[viii] Acetylene = a.cety.lene (sounds like cetiline; stress on 'ce')

[ix] Me (as M.E.), Et (as et), Pr (as P.R.) and Bu/But (as B.U. or But, 'u' as in you)

[x] Bu^t/t-Bu (*t*-butyl, *tert*.butyl, Bu = butyl group).
For further examples vide infra.

More examples of application of grammar

[Prefix = P.F., root name= R.N. and Suf. = suffix, Multiplying prefix = M.P.]

[i] CH_3Cl = chloro + meth~~a~~ + ane = chloromethane.
P.F. R.N. Suf.

[ii] CH_2Cl_2 = di.chloro + meth~~a~~ + ane = dichloromethane
M.P. P.F. R.N. Suf.

[iii] Penta = five (5),
(a) PCl_5 = phosphorus ('us' not 'ous'), pentachloride ('a' stays before 'c')
(b) P_2O_5 = phosphorus pentoxide (pent~~a~~+oxide, 'a' is dropped as 'o' follows)

[iv] Mono = one (1); CO = carbon monoxide (mon~~o~~ + oxide, 1st 'o' is dropped)

[v] Tetra = four (4); OsO_4 = Osmium tetroxide (tetr~~a~~ + oxide, 'a' is dropped)

[vi] CH_3—CHO = ethanal (Eth~~a~~ + an~~e~~ + al; 'a' and 'e' are dropped)
C_2 sat aldehyde

[vii] CH_3—COOH = ethanoic acid (Eth~~a~~ + an~~e~~ + oic; 'a' and 'e' are dropped)
C_2 sat. acid

[viii] CH_3—CH_2—CH_2—CHO = butanal (But~~a~~ + an~~e~~ + al; 'a' and 'e' are dropped).
It should be noted that when the vowel sound is missing (as in multiplying prefixes like, di, tri, etc., pronounced as dai, trai, the 'i' will not be dropped, e.g., trioic acid, 'ioi' stays; dioxide, 'io' stays) it should not be cancelled. Moreover, a few prefixes like *cyclo*, hydro, etc. do not drop their last 'o' in most of the cases, e.g., hydroanthracene; iodoethane, chloroacetone, etc. but hydracids are accepted.

[ix] The formula of a compound is written in English alphabet and read or spoken in English alphabet; moreover, the respective number of atoms in the formula is expressed in Arabic numerals (viz. 1, 2, 3, etc) and read or spoken as such, e.g., H_2SO_4 (read H-two-S-O- four and *not* like hydrogen-two-sulphur-oxygen-four), $C_6H_{12}O_6$, glucose (read C-six-H-twelve-O-six, and *not* like carbon-six-hydrogen-twelve-oxygen-six), HNO_3 (read HNO-three). Another point is to be noted. During showing the functional groups (atoms or group) always a dash is used before the symobls. This is not the minus sign of arithmatics. Thus —NO_2 is nitro group (not to be articulated as minus NO_2 group), —COOH is C—O—O—H group (or C—O-two–H group), not minus COOH or —CO_2H group or KOOH group.

6. Nomenclature

The organic compounds must be named to determine their identity. One method is the common (or trivial) name which is derived from several factors but the more recent method gives more clear picture of the given compound regarding its molecular architecture. The common name may be short and easy to remember but the latter method called IUPAC is very helpful to draw the structure of the molecule from the given name. If the IUPAC name of a molecule is complex and very long, the common name of the compound is accepted in IUPAC system of nomenclature as the parent compound and the substituents are indicated in proper locations with the respective symbols. The following examples are illustrative.

Examples

	Common name	Structure	IUPAC name	Remarks
[i]	Pivalic acid	$(CH_3)_3C-COOH$	Dimethylpropanoic acid	From IUPAC name the structure can be readily drawn and some chemical properties can be predicted.
[ii]	Neo-pentane	$(CH_3)_3C-CH_3$	Dimethylpropane	do.
[iii]	Amyl alcohol	$Me(CH_2)_3-CH_2OH$	Pentan-1-ol	do. (1°-alcohol)
[iv]	Oil of niobe	$C_6H_5-CO_2-CH_3$	Methyl benzoate	do. (ester)
[v]	Benzene	C_6H_6	Benzene	Accepted in IUPAC as the parent compound
[vi]	Toluene	$C_6H_5-CH_3$	Methylbenzene	do.

7. Anatomy of IUPAC Name of the Compounds

The following dissection of the IUPAC name will enable the beginners to understand the purpose and importance of each part of the name. This knowledge will be helpful to understanding the molecule from the name as well as writing the name correctly from the given structure.

Example [A]

The structure of the compound [A] is given below:

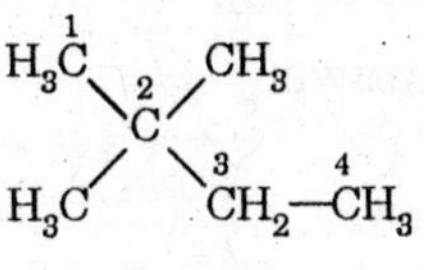

[A]

This is, 2, 2-Dimethylbutane (*pr.* 2, 2-Di.methyl.butane)

Anatomy of [A]

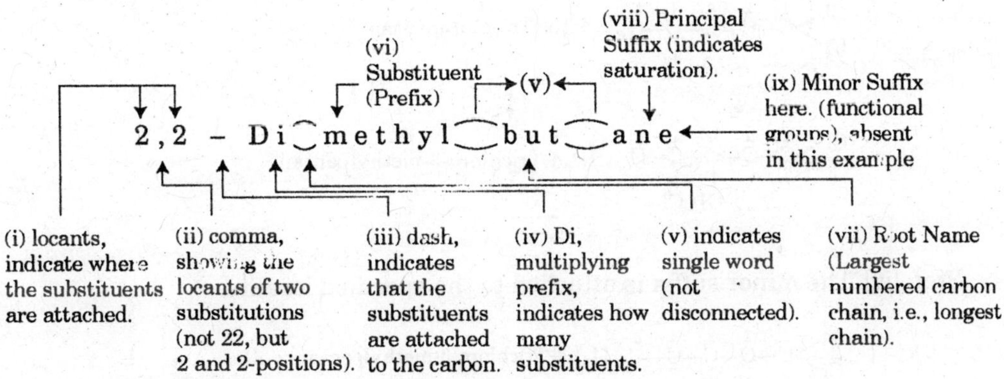

Notes: [a] **When the name of the compound is the first word, the first letter will be in higher case (i.e., capital letter) but in the middle of the sentence the first letter will be in lower case (i.e., small letters).**

[b] **The root name and the principal suffix must be present in all IUPAC nomenclature unless trivial name is accepted by IUPAC.**

Example [B]

The anatomy is shown below (of B)

1,1,4–Trichloro–3–methylpentan–2–ol [B]; IUPAC name.

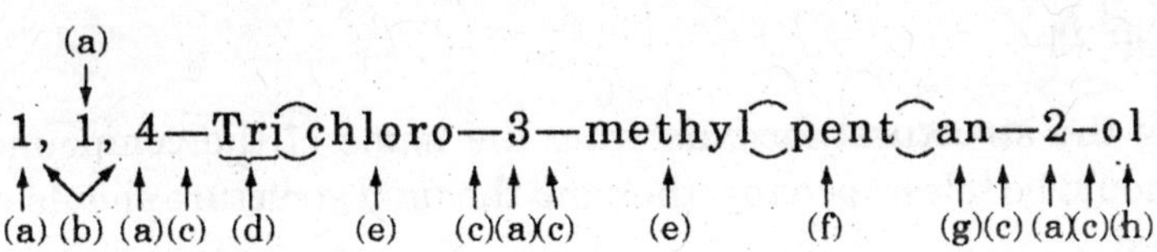

(The ⁀ sign has been explained in Ex. [A])

Legends: (a) Locant (s); (b) comma; (c) dash [in the middle of a name before and after the locant (s)]; (d) multiplying prefix (M.P.); (e) name(s) of the substituent(s) following the order of English alphabet (Prefix, coming before the root name and connected it without the dash); (f) root name (longest chains of carbon atoms furnish the root name); (g) principal (or major) suffix (attached directly to the root name following it), (h) minor suffix (indicates the functional group given the topmost priority (vide infra). Other functional group(s), if present, will be attached to the root name as prefix(es) with minor priority.

[a] Drawing the structural formula from the above nomenclature of [B].

Step [i] The root name is detected and the *skeletal chain* is drawn.
So, pent = C–C–C–C–C (This is 'f')
an (i.e. ane), C–C–C–C–C (This is 'g'), no unsaturation.

Step [ii] Locants are given $\overset{1}{C}-\overset{2}{C}-\overset{3}{C}-\overset{4}{C}-\overset{5}{C}$ (The carbons are marked by Arabic numerals, i.e., locants)

Step [iii] The prefixes are attached to the specified locants.

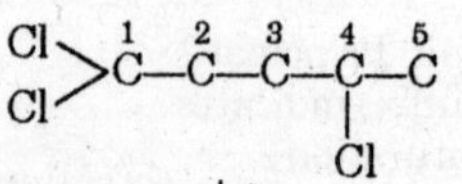

1,1,4-Trichloropentan

↓

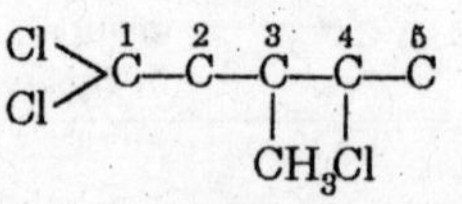

1,1,4-Trichloro-3-methylpentan

Step [iv] The minor suffix is attached to the specified locants.

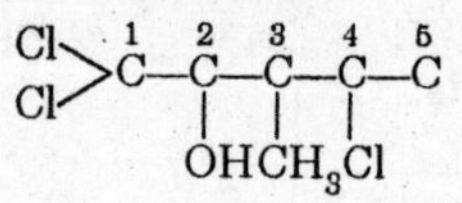

(1,1,4-Trichloro-3-methylpentan-2-ol)

Step [v] The vacant bonds of carbon are satisfied by H atoms.

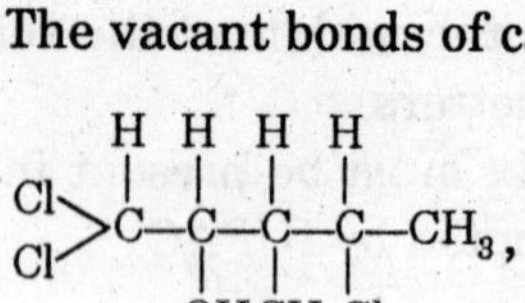

, Final structure in 'long-hand'.

and, 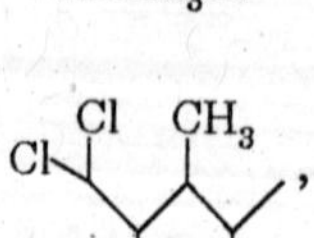 , Final structure in 'short-hand'.

Note: While writing the structural formula from the name of the compound all these steps should be telescoped together and the final structure should come in a single step.

[b] The following points are important

[i] The root name will be derived from the *longest carbon chain*. Any hydrogen atom when replaced by a group or atom will be called substitution and such group or atom will be called *substituents* on the involved carbon.

[ii] Arabic numerals will be used to denote locants. Thus 1,2,3, ..., etc. are used, not Roman numerals like I, II, III, etc. or any other numerals.

[iii] The prefixes will be arranged in the English alphabetic order and attached to the root name in the substituted molecule. The multiplying prefixes will be ignored to determine the alphabetic order. Thus methyl (i.e., 'M') comes after chloro (i.e., Cl) which comes after bromo (i.e., B), but all these substitutents come before nitro (i.e., N). Dimethyl (m considered, not d), and trinitro (n considered, not t) will be arranged as dimethyl trinitro compound but trichloro (i.e., c, not t) comes before dimethyl as trichlorodimethyl compound.

[iv] Smaller alkyl groups (C_nH_{2n-1}) are called branch of the main chain (root name).

[v] When more than one locants are present, then their sum is taken from all possible directions. The smallest sum will determine the locants of the substituents. This is known as the 'Rule of Lowest Sum'. If there is only one substituent, then also this rule is applied.

[vi] When the sum of the locants becomes same in both directions of numbering the carbons on the main chain, usually that name which starts with the smallest number will be considered as the correct one. This is arbitrarily taken as '4s' principle (same sum start small = 4's') for the sake of simplicity in this book used as a mnemonic.

[vii] The carbons on the branching may be marked as either 1′, 1″, ..., 2′, 2″, or 1, 2, ..., etc. as required and the multiplying prefixes be termed as *bis*, *tris*, ..., etc. in case of complex substituents (e.g., *bis*-dimethylamino group, but diamino group).

[viii] The most highly branched chain is followed when more than one route of the chain have the same chain-length (i.e., same root name).

[ix] In a very rare occasion when several *same* substituents occur in the chain, then that name which starts with the lowest number (or appearance of the lowest number in the series) will be considered as correct. (Rule of lowest number).

[x] Location designation by α, β, etc. and *ortho*, *meta*, etc. are not accepted in IUPAC systems of nomenclature.

Examples

[a] $Me_2CH—CH_2—CH(Et)Me$ ≡ 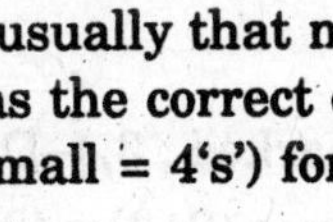

(C_8H_{18})

***Routes*:**

abc = 3-carbon chain }
fegh = 4- " " } not longest chains.
abdef = 5- " " } Rejected.
cbdef = 5- " " }

abdegh or cbdegh } = 6- " " Hexane } Correct choice of root name

If marked by Arabic ≡ 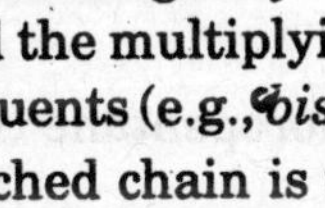

numerals in this way,
the name is 2,4-Dimethylhexane ... (I)

If marked by Arabic numerals from the opposite end, 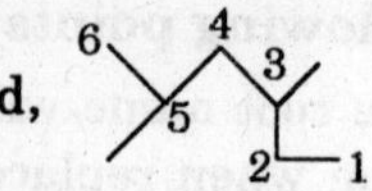

the name is 3,5-Dimethylhexane ...(II)
Rule of lowest sum is applied now. (addition of locants).
In I, 2+4 =6, and in II, 3+5=8, since 6 < 8, the name I is correct.

[b] $CH_3—C(CH_3)(Cl)—CH_3$ ≡ 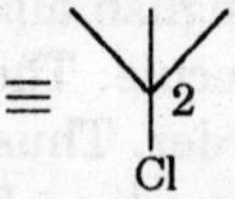
= 2-Chloromethylpropane (methyl needs no locant because in propane this is the only position open for methyl).

[c] $CH_3—CH(CH_3)—CH_2Cl$ ≡ 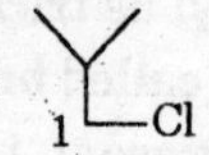
= 1–Chloromethylpropane (methyl needs no locant)

[d] 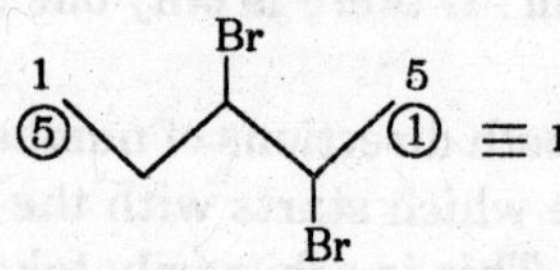 ≡ marked with circle; 2,3–Dibromopentane ... (I)

unmarked ; 3,4–Dibromopentane ... (II)

I is correct nomenclature since (2+3) < (3+4) (Rule of lowest sum)

[e] 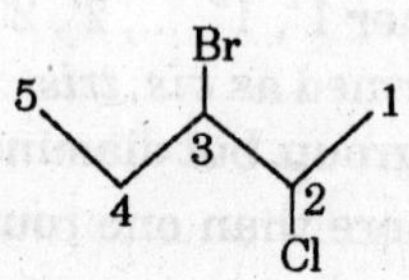 ≡ 3-Bromo-2-chloropentane (B comes before C, alphabetic order); not 3-Bromo-4-chloropentane

[f] 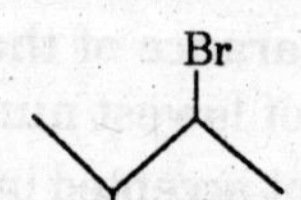 may be of four names,

In each name [i] through [iv] the sum of locants are same (2+3). Hence proposed '4s' is applied (vide vi)

[i] 2-Chloro-3-bromobutane (wrong; violets alphabetic order)
[ii] 3-Bromo-2-chlorobutane (wrong; starting bigger locant)
[iii] 2-Bromo-3-chlorobutane (right; follows '4s')
[iv] 3-Chloro-2-bromobutane (wrong; violates alphabetic order)

[**Note**: Priority of alphabetic order > priority of locants].

So, 2-chloro-3-methylbutane is the correct name of 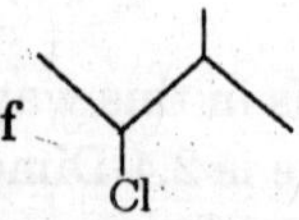

Example [C]

The common name of [C] is *p*, *p*′–dichlorodiphenyltrichloroethane which is commercially known as DDT. The IUPAC nomenclature of DDT is 1,1,1–trichloro–2,2,-*bis* (4′–chlorophenyl)–ethane [C]. The dissection of [C] is shown below.

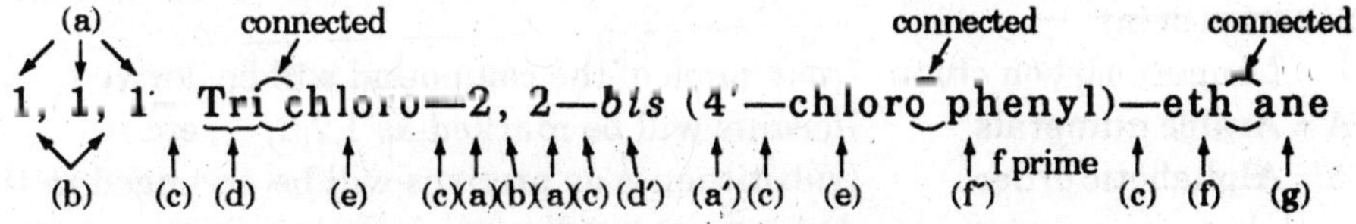

Legends: (Vide example [B]), (d′) is multiplying prefix of complex substituents indicating two such groups, prime on Arabic numerals indicates locant on the substituents not on the main carbon chain.

[a] Drawing the structural formula of [C] from the IUPAC name.

[i] *Root name* (f) *etha*; so chain of two carbons (C—C), marked (C^1—C^2).

[ii] *Suffix* (g), (major) is *ane*; so this is saturated, thus (eth a + ane) = ethane (grammar).

[iii] *Prefix* is chloro (e, on the *main chain*), multiplying prefix is *tri* (d, simple substituent), locants are 1,1,1 (a, separated by comma, pronunciation one, one, one; not as one hundred and eleven, 111). Therefore, the partial structure of [C] is $\left(Cl_3\overset{1}{C}—\overset{2}{C}\right)$.

[iv] The complex substituents on the main carbon chain (root name) is 4′–chlorophenyl. This is (Cl— 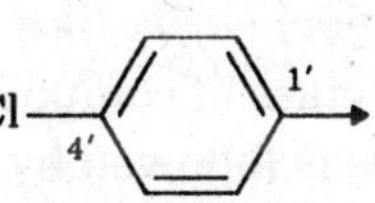 group), two such complex groups (*bis*, d′) are on 2,2–locants(a) on the ethane.

[v] So the complete structure of [C] is 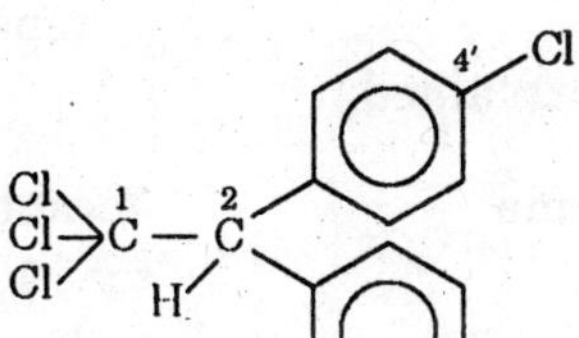 [C].

[b] From the foregoing examples [A], [B] and [C] it is evident that only two items like *root name* and *major suffix* (like ane, ene, yne or enyne) can furnish the complete IUPAC nomenclature of a compound. But some nomenclatures may require many more items (found a maximum of twelve items, excluding parentheses). In the IUPAC nomenclature these items should be checked to determine the correctness of the name. For this purpose the following mnemonic (method for easy remembering) may be very useful.

8. Mnemonic

1 2 3 4 5 6 7 8 9 10 11 12
LAAM, LoRoo, SuPre, CoDa, VoNo (12-items)

[a] Pronounced as, Laam, Loroo, Supre, Coda, Vono.

[b] Explanation of (a)

1. L = **Longest carbon chain** — (*root name* of the compound will be derived)
2. A = **Arabic numerals** — (*locants* will be marked as 1,2,3, ..., etc.)
3. A = **Alphabetic order** — (substituents as prefixes will be arranged in this order, e.g., oxa will come before oxo).
4. M = **Multiplying prefix** — [number of same substituents when *simple* is shown as di, tri, tetra, ..., etc. But *complex* substituents are shown as *bis* (i.e., 2), tris (i.e., 3), etc. Multiplying prefixes are *not* separated from the corresponding prefixes by a dash (i.e., hyphen). They appear as a single word (e.g., dichloro, tetranitro, etc.)]

   LAAM is explained.

5. Lo = **Lowest locant** — (locant should have lowest number when single, and the lowest sum when more than one locants are present).
6. Roo = **Root** name — (determined from the longest carbon chain).

   LoRoo is explained.

7. Su = **Suffix** — [fixed with the root name just after it as the principal (major) *suffix* like ane, ene, yne or enyne indicating the class of compound like saturated or unsaturated. This is followed by *minor suffix* (one only) which represents the functional group.]
8. Pre = **Prefix** — (fixed with the root name just before it, locants will be separated by a dash).

   SuPre is explained.

9. Co = **Comma** — [locants of the *same* prefixes are separated by comma, e.g., 1,1,1–means *same* substitutent occuring thrice and articulated as one, one, one (not like one hundred and eleven as 1,1,1 ≠ 111.)]
10. Da = **Dash** — (represents direct connection with the carbon of the main chain. Locants are connected to the prefixes by dash, single at the beginning and double between two prefixes, e.g., 2-bromo-3-chlorobutane).

    CoDa is explained.

11. Vo = **Vowels** — (when *two* vowels come together, the first vowel is dropped in most of the names. Vowel-sounding consonants like y in 'yne' is also treated as vowel. This is the grammar of nomenclature).

12. No = **Numero** =**Number** [when the sum of the locants becomes same in all directions, then that nomenclature which starts with the lowest numbers (locant) will be considered. This is rule of lowest number (proposed mnemonic as '4s').].

[c] If the vowels do not sound like vowels, the grammar of cancellation of the first vowel is not obeyed, e.g., d*i*ol, tr*ie*ne. Some prefixes also retain their last vowels, e.g., cyclo, oxo, ox*a*, tetra, etc. viz. cyclooctatetraene, oxoaza, oxaene. The locants in between the two vowels under consideration is ignored. On the contrary, if the consonant sounds like a vowel, then its preceding vowel is generally cancelled, e.g., (enę+yne) = enyne.

[d] *Application* of the mnemonic :

[i] The given structure of the compound is [A]. 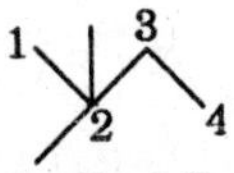 = [A]

[ii] Correct IUPAC nomenclature (written) = 2, 2-Dimethylbutane.

[iii] Pronounced as (spoken) = Two.two.di.methyl.butane.

[iv] The following expressoins (written and/or spoken) are wrong (*vide mnemonic to identify the errors*).

1. 2,2-Butanedimethyl
2. Dimethyl-2, 2-butane
3. 2,2-Methyldibutane
4. Dimethylbutane-2,2
5. 2 2-Dimethylbutane
6. 2,2-Dimethylbutan
7. 2,2-Dimethyl-butane
8. 2,2 Dimethylbutane
9. 2,2-Bismethylbutane
10. 3,3-Dimethylbutane
11. 2,2-Di.methyl.butane (when written)

[e] **Expansion:** (i) A compound when saturated will be suffixed as 'ane'. The unsaturated compounds may be 'ene' (contain C=C) or 'yne' (contain C≡C). Since a compound cannot be saturated and unsaturated at the same time, so 'ane' will never be combined with 'ene' or 'yne'. But the unsaturation 'ene' or 'yne' may be combined like 'enyne' (as in C=C—C≡C). The radicals (as substituents) will be alkyl (from alkane), alkenyl (from alkene) and alkynyl (from alkyne) (vide infra).

Examples

1. 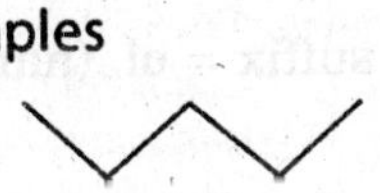 = Pentane.
2.  = 2-Pentene or Pent-2-ene (*Not* 2-Pentanene or 3-Pentene).
3. 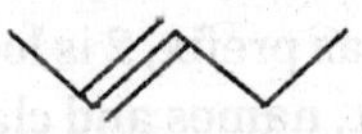 = 2-Pentyne or Pent-2-yne (*Not* 2-Pentanyne).
4. 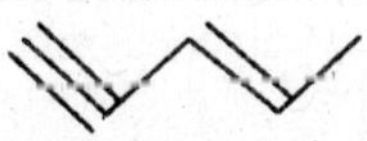 = Pent–3–enyne (enyne, both C=C and C≡C exist).

(*Not* Pent-3-en-1-yne), locant of yne is not required (because 3-ene keeps only C_1 for yne).

5. $CH_2{=}CH_2$ Ethene $\xrightarrow{[-H]}$ $CH_2{=}CH—$ (ethenyl radical or group).

6. CH≡CH Ethyne $\xrightarrow{[-H]}$ CH≡C— (ethynyl radical or group).

7. 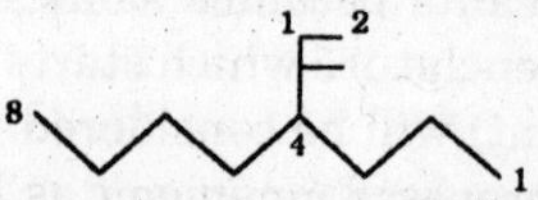 = 4-Ethenyloctane (octane longest chain; ethenyl is prefix, so 'ene' can stay as a side chain but not on the main chain). *Not* 3-propylhept-1-ene or 5-ethenyloctane (since 4<5, Rule of lowest number).

8. 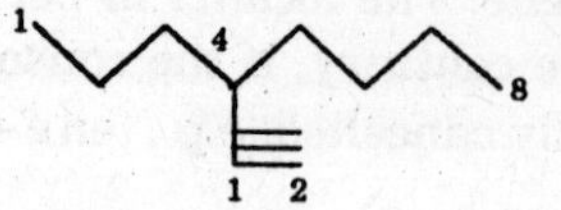 = 4-Ethynyloctane

↑ Smaller locant; Prefix; root name (saturated)

(not 5-ethynyloctane). *Not* 3-propylhept-1-yne.

9. 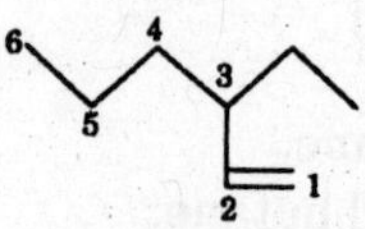 = 3-Ethylhex-1-ene (longest chain holding the functional group, "ene" is taken).

[ii] From the foregoing description it is evident that the prefix and the suffix are related to the root name, and all of them together with the locants reveal the molecular architecture. The same functional group may be the prefix in one compound while it may be the suffix in another compound. During the change the functional group generally changes its expressions. When only one functional group is present then it is made the minor suffix with a very few exceptions (e.g., NO_2, nitro group, is always used as a prefix, never suffix.) But when two or more functional groups are present in the molecule then that group having *maximum priority* wil be suffix and the remaining functional groups will be prefix (vide infra). It should be remembered in this connection that even a single functional group may be used as a prefix with the 'ane', 'ene', 'yne' or 'enyne' as the major suffix.

Examples

Functional group —OH (not in —C(=Q)—OH); prefix = hydroxy; suffix = ol. (name = hydroxyl group). Q = heteroatom (eg. N, O, S, P etc.)

Compound is CH_3—CH(OH)—CH_3 = Propan-2-ol (—OH as minor suffix and ane as major suffix)

=2-Hydroxypropane (—OH as prefix, 2 is locant)

[iii] The following table illustrates the functional group, prefix, suffix, names and class of compound which is to be remembered by the beginners.

9. List of Several Functional Groups and Names in IUPAC Nomenclature (FG = Functional Group)

Sl. No.	Functional Group	Class of compd. holding the FG	Formula of the FG	Name of FG when Prefix	Name of FG when Suffix
[i]	Hydroxyl	alcohol and phenol	–OH	hydroxy–	–ol
[ii]	Halogen	halide	–X (X = F, Cl, Br and I)	halo– (chloro, bromo, etc)	halide
[iii]	Oxa	ether/alkoxy	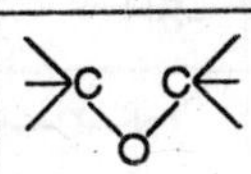	oxa–	alkoxyalkane (ether)
[iv]	Aldehyde	carbonyl	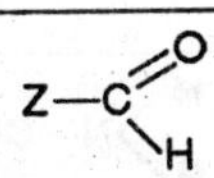 (Z = H or C)	oxo–, formyl–	–al
[v]	Keto	ketone/carbonyl	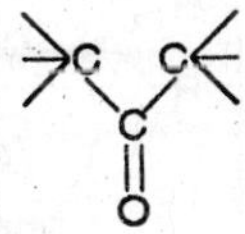	keto–, oxo– (cf. iv)	–one (*Pr.* own)
[vi]	Acid	acid, carboxy acid	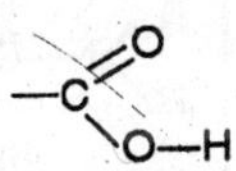	carboxy–	–oic acid
[vii]	Acid anhydride	anhydride	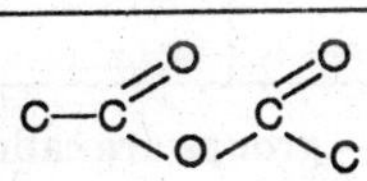		–oic acid anhydride
[viii]	Ester	ester	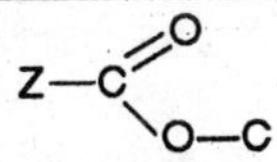 (Z = H or C)	alkoxycarbonyl or carbalkoxy	–oate (*Pr.* wet)
[ix]	Acid halide	carbonyl halide	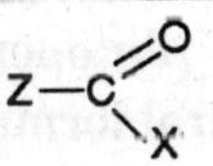 (Z=C; X=Hal)	haloformyl– or halocarbonyl–	–oyl halide
[x]	Acid amide	amide	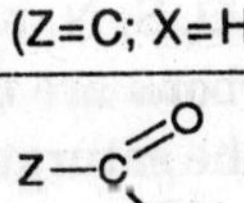	carbamoyl–	carbonamide
[xi]	Amino	amine-1° amine-2° amine-3°	$-NH_2$ (1°) –NHR (2°) (R≠H) $-NR_1R_2$(3°)	amino–	–amine

Sl. No.	Functional Group	Class of compd. holding the FG	Formula of the FG	Name of FG when Prefix	Name of FG when Suffix
[xii]	Cyano/nitrile	cyanide	–CN	cyano–	–nitrile
[xiii]	Nitro	nitro	$-NO_2$	nitro–	*Not* suffix
[xiv]	Nitroso	nitroso	–NO	nitroso–	*Not* suffix
[xv]	Sulphonic acid	sulphonic acid	$-SO_3H$	sulpho–	sulphonic acid
[xvi]	Sulphoxide	sulphoxide	$-C-\overset{O}{\overset{\Vert}{S}}-C-$ (S=O)	Sulphinyl–	*Not* suffix
[xvii]	Thiol	thiol	–SH	mercapto–	–thiol
[xviii]	Urea [*Pr.* you.ria]	urea	$>N-\overset{O}{\overset{\Vert}{C}}-N<$ (C=O)	ureido–	urea
[xix]	Monosubstituted urea	Substituted urea	$(H)(R)N-CO-N(H)(H)$	–	N-Substituted urea
[xx]	Disubstituted urea	do.	$(H)(R)N-CO-N(H)(R)$ (a) and $(R)(R)N-CO-NH_2$ (b)	–	[a] N,N′-disubstituted urea [b] N,N-disubstituted urea

Note: The $-CH_2-$, and $-CH_2-CH_2-$ groups are called methano and ethano, respectively. Also the $-\overset{|}{C}H-$, CH (methylidyne), $CH_3-CH\equiv$ (ethylidyne) are used.

10. Nomenclature of Hydrocarbons

Organic compounds made of only carbon and hydrogen are called hydrocarbons. They are either saturated (general formula C_nH_{2n+2} for open-chain hydrocarbons, and for monocyclic compounds C_nH_{2n}) or unsaturated (general formula C_nH_{2n} when only one C=C is present, and C_nH_{2n-2} when only one C≡C is present). Any element other than C and H is known as heteroatom in organic chemistry (e.g., O, S, N, halogens, etc).

The open chain saturated hydrocarbons are called alkane, the 'alk' is the generalised root name and 'ane' is the signature of the saturation (i.e., absence of C=C and C≡C). The unsaturated hydrocarbons are of two simple types, viz., alkene (contains C=C) with signature 'ene', the other type is alkyne (contains C≡C). The mixed types of unsaturated hydrocarbons are alkenynes (contain C=C plus C≡C). The multiplying prefixes like di, tri, etc. are used to denote the number of C=C and/or C≡C when present in the hydrocarbons.

When a hydrogen atom is removed from the carbon of the chain, the redical is obtained. When this vacant spot is filled in by an atom (not H) or group (functional), then this phenomenon is called *substitution* and the incoming group or atom is called *substituent* which dramatically changes the chemical, physical and physiological properties of the hydrocarbon. This is shown below.

[a] Saturated hydrocarbons

[i] $CH_4\ (C_nH_{2n+2}) \xrightarrow[\text{[-ane] [+yl]}]{[-H^\bullet]} CH_3^\bullet$;

methane (*n*=1) — *methyl radical*

$(C_nH_{2n+2}) \xrightarrow[\text{removed 'ane' added 'yl'}]{-H^\bullet} (C_nH_{2n+1})^\bullet$

alkane — alkyl group (radical)

[ii] $C_2H_6\ (n=2) \xrightarrow{\text{do.}} CH_3{-}CH_2^\bullet$

ethane — ethyl

[iii] $\overset{1}{CH_3}{>}\overset{2}{CH_2}$ (CH_3, CH_3) $\xrightarrow[C_1]{[-H^\bullet]\text{ from}} CH_3{-}CH_2{-}CH_2^\bullet$

Propane — *n*-propyl group

$\xrightarrow[C_2]{[-H^\bullet]\text{ from}} (CH_3)_2{>}CH^\bullet$

iso-propyl group

[iv] $C_{20}H_{42} \xrightarrow[\text{terminal C}]{[-H^\bullet]} C_{20}H_{41}^\bullet$

(*n*=20) — icosyl group

{eicosane – oldspelling
icosane – new spelling}

[v] $C_{100}H_{202} \xrightarrow[\text{terminal C}]{[-H^\bullet]\text{ from}} C_{100}H_{201}^\bullet$

(*n*=100)
hectane — *n*-hectyl group

[b] Unsaturated hydrocarbons (alkenes and alkynes)

[i] $CH_2{=}CH_2\ (C_nH_{2n}) \xrightarrow[\text{terminal 'e' dropped. 'yl' added}]{[-H^\bullet]} CH_2{=}CH^\bullet$; alkene $\xrightarrow[{[-e][+yl]}]{[-H^\bullet]}$ alkenyl

ethene (n=2) — ethenyl radical — (C_nH_{2n-1}) radical

[ii] $CH{\equiv}CH\ (C_nH_{2n-2}) \xrightarrow[\text{terminal 'e' replaced by 'yl'}]{[-H^\bullet]}$ ethynyl radical $(C_nH_{2n-3})^\bullet$;

ethyne (n = 2)

alkyne $\xrightarrow{[-H^\bullet]}$ alkynyl radical

11. Nomenclature of Alkanes

This is done stepwise as shown below.

[i] The longest chain is identified (*Roo*), [ii] The carbons are numbered by Arabic numerals (A), [iii] The side chain, if present as substituent, must be an alkyl group (because the alkane is under consideration). This side chain is often referred to as branching of the main chain. The locant of this branching is given the lowest number (*Lo*), [iv] The larger side chain will constitute the main chain (*Roo*), [v] The locant of the side chain (i.e., branching) will be connected to the name of the alkyl side chain by a dash (*Da*).

Examples

Taking penta as the root name, three methods of representation are shown.

[i] $CH_3{-}CH_2{-}CH_2{-}CH_2{-}CH_3 \rightarrow \overset{1}{C}{-}\overset{2}{C}{-}\overset{3}{C}{-}\overset{4}{C}{-}\overset{5}{C} \rightarrow$ 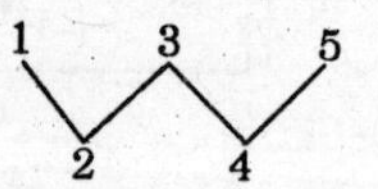 (carbons marked by Arabic numerals)

Pentane (no branching) 'Long-hand' drawing time consuming and clumsy.

Pentane 'Skeletal' formula, time consuming but clean.

Pentane 'Short-hand' formula, quick and clear.

[Five-carbon chain (Penta + ane) = Pentane (saturated hydrocarbon, 'ane')]

(Roo ↓ Penta; Suf ↓ ane; ↑ dropped)

[ii] $\overset{1}{CH_3}—\overset{2}{CH}(CH_3)—\overset{3}{CH_2}—\overset{4}{CH_2}—\overset{5}{CH_3}$ →

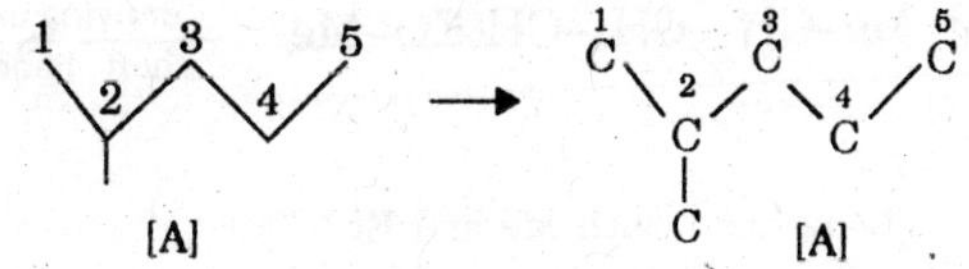

[A] [A]

Carbon chain with branching. The branch is CH_3 group (substituent)

2-Methylpentane

(location of Me) (dash) (prefix) (Roo) (Suf.)

Numbering from other end (by O) (Wrong numbering) →

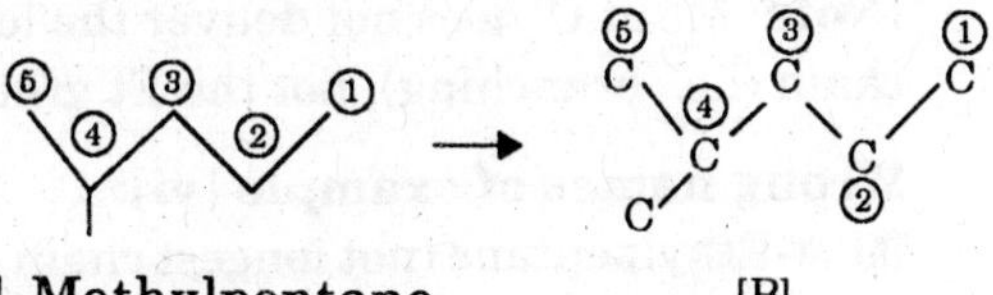

4-Methylpentane [B]

[B]

Since [A] has the locant 2 smaller than that of [B] (i.e., 4), the [A] is the correct name. So the locant should be smaller whenever the carbons of the molecules can be marked both ways.

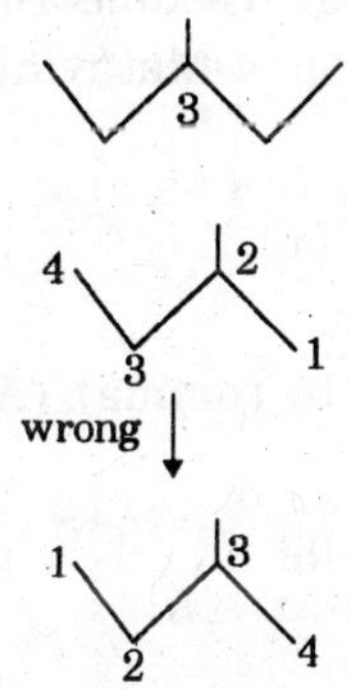

[iii] This is 3-methylpentane (no problem in numbering, note locant, dash, prefix with no dash, connected to root name followed by suffix obeying the grammar).

[iv] This is methylbutane [note methyl is substituent, 'm' is not capital letter because the name is in the middle (or end) of a sentence, and the name *does not require* the locant of the Me group]. 2-Methylbutane [no need of 2 (locant), since the Me at C_1 or C_4 will make the five carbon chain (pentane), and Me at C_3 will make this carbon marked as 2, because locants must be smaller. Note that if the correct structural formula can be drawn from the locantless name, then there is no need of the locant.]

[v] Methylpropane → (Drawing)

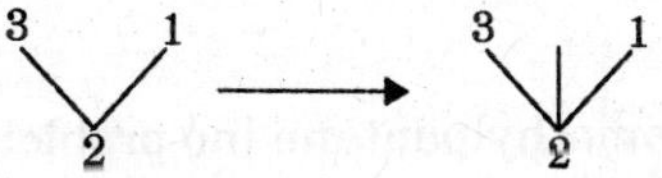

(No locant, C_2 is the *only* position where the Me can be fixed).

(No locant) Propane

Dimethylpropane

M.P. Pre Root Suf → (Drawing)

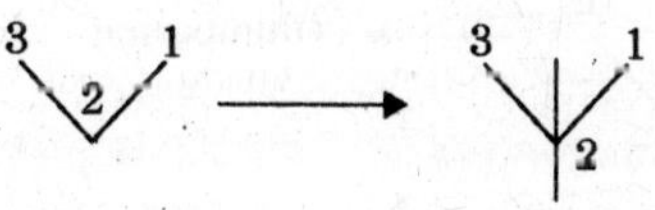

(do.)

(No dash, no locant)

(2,2-Dimethylpropane is wrong)

[vi] Me–CH_2–CH_2–CH(Et)–Me $\xrightarrow[\text{short-hand}]{\text{drawing in}}$ 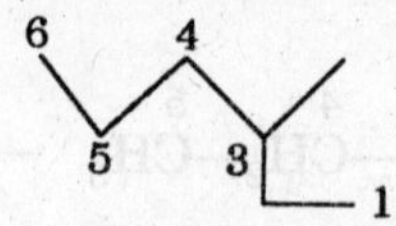 Clean picture of the given compound

('Long-hand' with Me and Et. Very confusing and crowded, better the 'short-hand' structure is drawn).

3-Methylhexane

[**Note**: Me at C_3 does not deliver the longest chain but Et group does, so Me is the side chain (i.e., branching), not the Et group in the whole molecule.]

Wrong names of example [vi]:

[a] 4-Ethylpentane (not longest chain).
[b] 2-Ethylpentane (do.)
[c] 4-Methylhexane (larger locant)
[d] 4 Methylhexane (no dash)
[e] 4-Methyl-hexane (dash between methyl and hexane)
[f] Hexane-4-methyl (prefix is displaced)
[g] 4-Methylhexaane (grammer disobeyed)

[vii] [a] 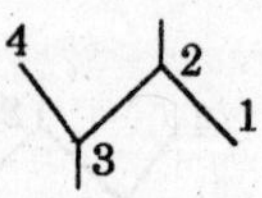 = 2,3-Dimethylbutane–[A]; Dimethylbutane needs locants (separated by comma). (A≠B, they are isomers). But methylbutane does not require locant.

[b] 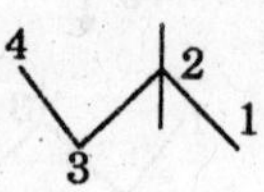 = 2,2-Dimethylbutane [B]; [Methylbutane does not require locant].

[viii] [a] 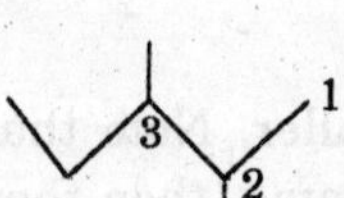 = 2,3-Dimethylpentane

[not 3,4-dimethylpentane, since (2+3)< (3+4); rule of lowest sum.]

[b] 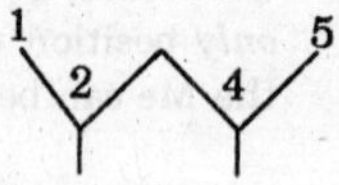 = 2,4-Dimethylpentane [no problem for fixing locants].

[ix] [a] 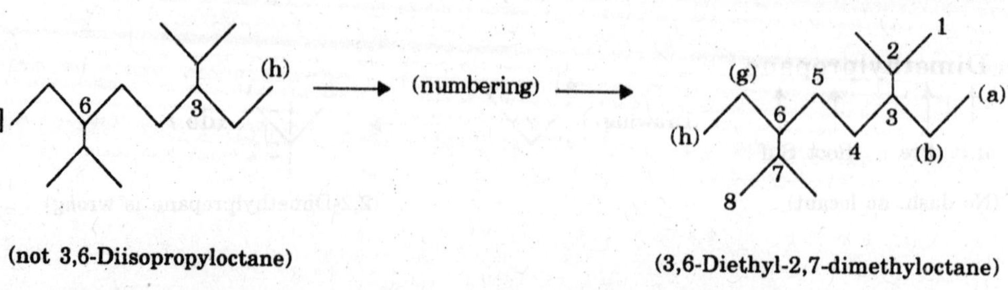

(not 3,6-Diisopropyloctane) (3,6-Diethyl-2,7-dimethyloctane)

['e' comes before 'm'; 'D' in di (M.P.), is ignored. Alphabetic order].

When the chain length is same in all forms of numbering, that route holding the maximum number of side chain must be considered to be correct. Thus a,b, ... , g, h (8-carbon chain) is ignored. The 1,2, ..., 7,8 route is correct.

[x] When same substituent occurs several times in a molecule at different locations (i.e., locants), then the 'rule of lowest sum' (i.e., the sum of all the locants should the smallest) is not followed. On the contrary, the 'rule of lowest number' is applied to determine the correct name. This rule of lowest number states that when locants of two alternative nomenclatures are compared with each other, then that *series* of locants which has the first appearance of lower number will be correct[21]. The following examples will illustrate the point.

[a]

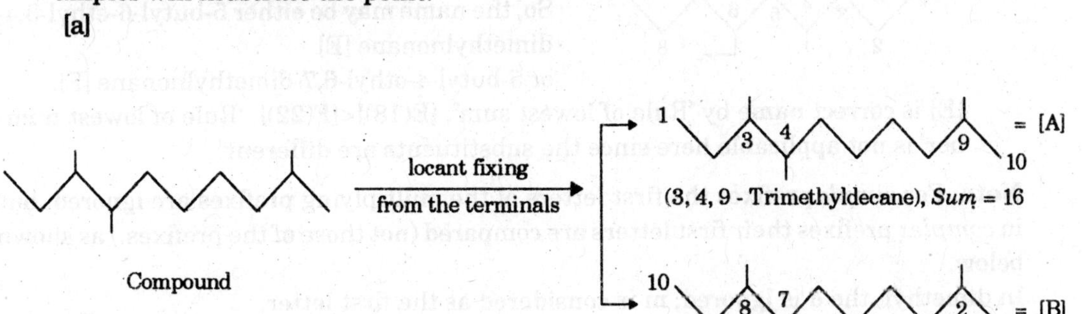

(3, 4, 9 - Trimethyldecane), *Sum* = 16

(2, 7, 8 - Trimethyldecane), *Sum* = 17

In [A], the sum of locants = 16.
In [B], the sum of locants = 17.
So it was anticipated that [A] should be correct (Rule of lowest sum) but actually [B] is accepted as the correct nomenclature by the application of Rule of lowest number, [B] starts with 2 while [A] starts with 3.

[b]

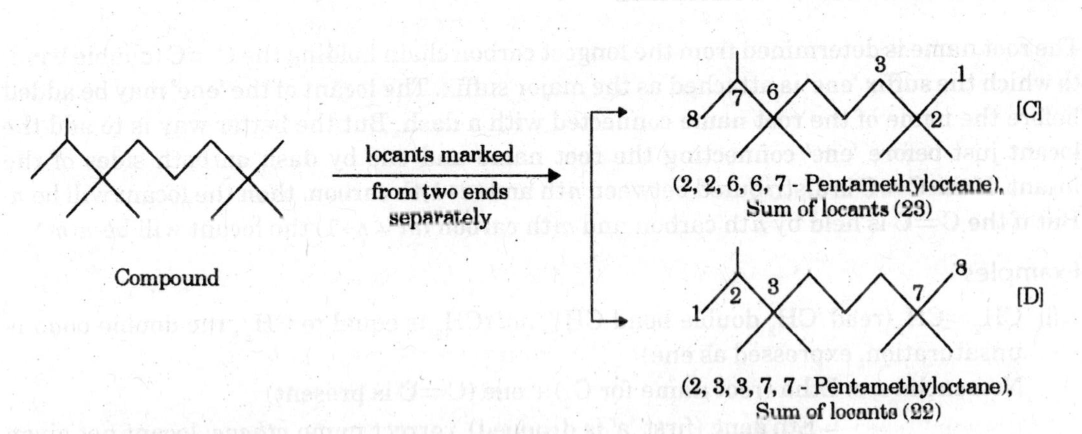

(2, 2, 6, 6, 7 - Pentamethyloctane), Sum of locants (23)

(2, 3, 3, 7, 7 - Pentamethyloctane), Sum of locants (22)

[C] is correct as the series has second locant (2) lower than second (3) in the series [D]. The first locants (2) are same in both the series, hence the second locants are compared which settled the nomenclature.

[c] At first the name of the prefixes are arranged according to the English alphabet (of English Dictionary) followed by their respective multiplying prefixes, then putting the corresponding locants separated by comma and connected by dash.

Example

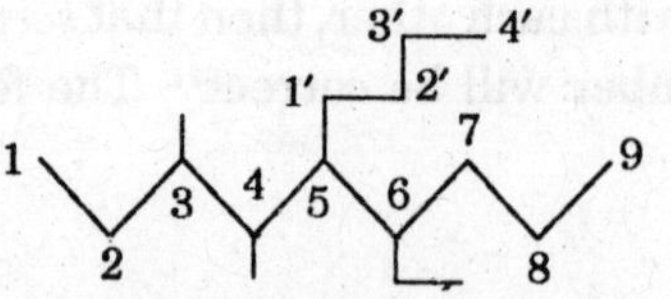

Prefixes (Substituents)
Dimethyl (m taken, not D)
Butyl =B–1st comes
Ethyl = E–2nd comes
Dimethyl = M–3rd comes
So, the name may be either 5-butyl-6-ethyl-3,4-dimethylnonane [E]
or 5-butyl-4-ethyl-6,7-dimethylnonane [F].

[E] is correct name by 'Rule of lowest sum', [E(18)]<[F(22)]. 'Rule of lowest number' is not applicable here since the substituents are different.

Note: For *simple* prefixes the first letters of the multiplying prefixes are ignored, but in *complex* prefixes their first letters are compared (not those of the prefixes.) as shown below.
In dimethyl, the d is ignored, m is considered as the first letter.
e.g., [i] In trichloro, the t is ignored, c is considered as the first letter. In tetranitro, the t is ignored, n is considered as the first letter.
So, the arrangement will be trichloro-dimethyl-tetranitro (but when the name is the first word in a sentence, the first letter of the multiplying prefix will be capital).
[ii] In *complex* multiplying prefixes their first letter is considered for comparison. In dimethylamino, d is considered, not m as the first letter, (complex prefix). In *bis* (4′-chlorophenyl), b is considered, not c as the first letter (complex prefix).

12. Nomenclature of Alkenes[21]

The root name is determined from the longest carbon chain holding the C=C (double bond) to which the suffix 'ene' is attached as the major suffix. The locant of the 'ene' may be added before the name of the root name connected with a dash. But the better way is to add the locant just before 'ene' connecting the root name and ene by dash on both sides of the locant. If the C=C is distributed between *n*th and (*n*+1)th carbon, then the locant will be *n*. But if the C=C is held by *n*th carbon and *m*th carbon ($m \neq n+1$) the locant will be *n*(*m*).

Examples

[i] $CH_2=CH_2$ (read 'CH_2 double bond CH_2', *not* 'CH_2 is equal to CH_2', the double bond is unsaturation, expressed as ene)
Nomenclature: Etha (root name for C_2) + ene (C=C is present)
= Eth ~~a~~ene (first 'a' is dropped), correct name *ethene*, locant not given as it is unnecessary.

[ii] $CH_3—CH=CH_2 \rightarrow$ Propa (root name for C_3) + ene = Prop ~~a~~ene = Propene
Locant of C=C not needed, as $\overset{3}{C}H_3—\overset{2}{C}H=\overset{1}{C}H_2$ is $\overset{1}{C}H_2=\overset{2}{C}H—\overset{3}{C}H_3$.

[iii] [a] $\overset{4}{C}H_3-\overset{3}{C}H_2-\overset{2}{C}H=\overset{1}{C}H_2$ and $\overset{1}{C}H_3-\overset{2}{C}H_2-\overset{3}{C}H=\overset{4}{C}H_2$,

(correct marking) (wrong marking)

This gives 1-ene. This gives 3-ene.

Locant necessary, this is either 1-butene or but-1-ene (better).

[b] $\overset{1}{C}H_3-\overset{2}{C}H=\overset{3}{C}H-\overset{4}{C}H_3$ $\xrightarrow[\text{stereoisomers}]{\text{two}}$

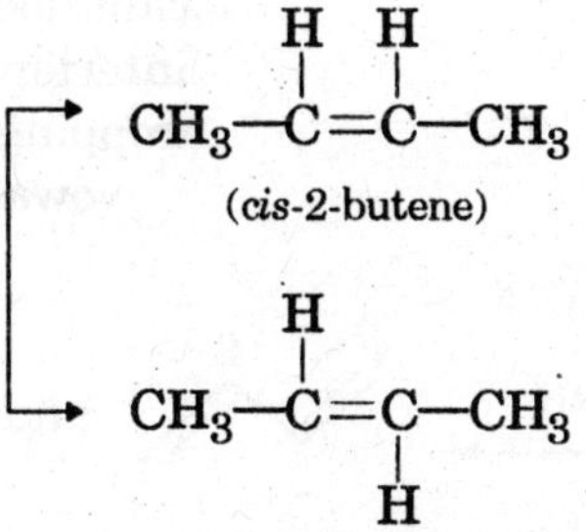

2-Butene or But-2-ene

(*cis*-2-butene) (Two H on the *same* side of C=C)

(*trans*-2-butene) (Two H on the *opposite* side of C=C)

[c] and *iso*-butylene, $CH_3-\overset{CH_3}{\overset{|}{C}}=CH_2$.

(not butene); (methylpropene, no locant).

Note: Butene, (C_4H_8), means a chain of four carbons having a double bond inside the chain. Depending on the position of C=C, two compounds are possible, viz., 1-butene (the C=C is between C_1 and C_2) and 2-butene (C=C is between C_2 and C_3, n=2, n+1=3). Moreover, 2-Butene can represent the *cis* (the H's on C=C are on the same side, i.e., close) and the *trans* (the H's on C=C are on opposite side, i.e., distant). So,butene can represent three isomers, i.e., same chemical formula but different compounds due to different atomic arrangements in the molecule, viz., 1-butene, *trans*-2-butene and *cis*-2-butene.

[iv] [a] 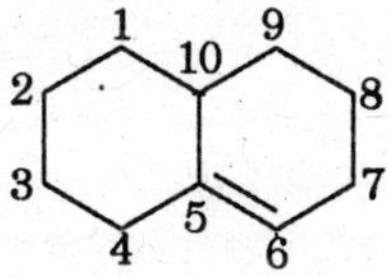 and 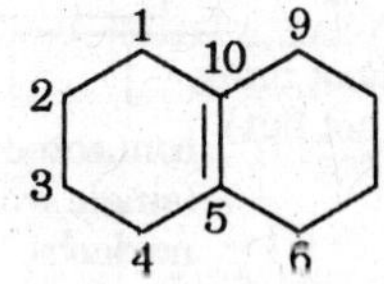

This is '5-ene'

[The C=C is distributed between C_5 and C_6 .So only 5 is the locant of C=C.]

This is '5(10)-ene'

[The C=C is distributed between C_5 and C_{10}; i.e., 5=n but 10≠n+1. So both 5 and 10 are stated as 5(10) as the locants of the C=C.]

[b] When more than one (C=C) are present in the main chain, then appropriate multiplying prefixes (di, tri, etc as the case may be) are added before the 'ene' (without the dash) with proper locants by the method shown before. Generally the vowel of the multiplying prefix just before 'e' of ene is not dropped (e.g., diene, triene, tetraene, etc. retaining 'ie', 'ae' in the spelling).

Examples

 —2H→  and 

[i] C_4H_{10},(C_nH_{2n+2}) Butane | C_4H_8,(C_nH_{2n}) But–1–ene (Buta-1-ene) locant does not interfere in dropping the vowel | C_4H_8,(C_nH_{2n}) But–2–ene (2-Butaene) (or 2-Buta-2-ene)

Butane —4H→ 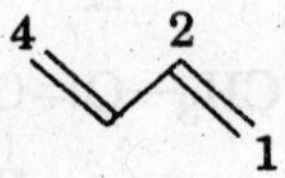 and 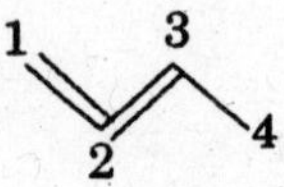 [B]

C_4H_6,(C_nH_{2n-2})

But-2-ene —2H→ is $\overset{1}{C}H_2=\overset{2}{C}=\overset{3}{C}H-\overset{4}{C}H_3$

is 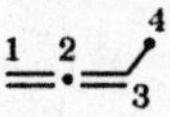 [A]

$C_4H_6(C_nH_{2n-2})$

Buta-1,3-diene
('a' stays as 'd' follows;
The 1st C=C is between C_1 and C_2;
the 2nd C=C is between C_3 and C_4;
dash before and after the locants separated by comma.)

Buta-1,2-diene
(The 1st C=C is at C_1 and C_2,
the 2nd C=C is between C_2 and C_3,
The C_2 carbon is sp hybridised,
so C_1—C_2— C_3 chain is linear.
The short-hand structure is shown in [A]

[ii] Anatomy of 1,3-butadiene or buta-1,3-diene

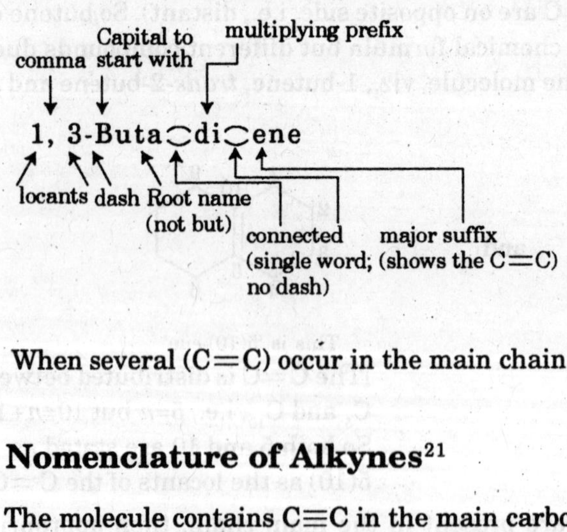

[iii] When several (C=C) occur in the main chain, the compound is termed 'polyene'.

13. Nomenclature of Alkynes[21]

[a] The molecule contains C≡C in the main carbon chain (read as 'C triple bond C') and major suffix is 'yne'. In alk-n-yne, the location of the C≡C is between C_n and C_{n+1}. The 'y' in yne is taken as the vowel following the grammar. When in the molecule both ene and yne are present, the ene (alphabate e) comes before the yne and the combination is en eyne = enyne. The general formula C_nH_{2n-2} stands for diene as well as yne. The

functional group, if present, is added as minor suffix after the yne with the locant of the functional group.

Examples

[i] CH≡CH = Ethyne (C_2; Eth ~~a~~ + yne); H—•≡•—H; (C_2H_2)

[ii] CH≡C—CH_3, Propyne (C_3; Prop ~~a~~+yne); H—•≡•—CH_3; (C_3H_4)

[iii]  ; 1–Butyne (C≡C is distributed between C_1 and C_2); (C_4H_6); (H—•≡•—/)

[iv] 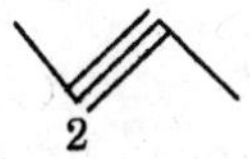 ; 2–Butyne (C≡C is existing between C_2 and C_3); (C_4H_6), (—•≡•—)

[b] When more than one yne are present, the corresponding multiplying prefixes are added to the yne (without dash) and the appropriate locants are given (separated by comma) and connected.to the main chain with dash.

Examples

[i]  i.e., HC≡C—C≡CH, is butadiyne (locant unnecessary as this is the only structure of butadiyne and 'a' in buta stays because it is followed by 'd').

[ii] 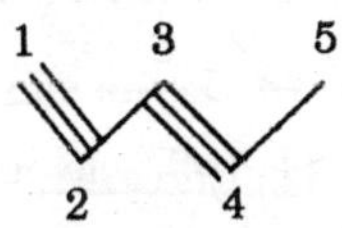

[A]

= Penta-1,3-diyne [A] the locants are necessary to distinguished it from penta-1,4-diyne [B]. The [A] is not penta-2,4-diyne, obeying the 'Rule of lowest sum'.

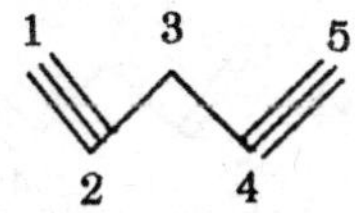

[B]

[iii]  This is hexatriyne. The locant is unnecessary because this compound has only one structure possible for the given name. [i] and [iii] are linear.

[c] When the molecule holds both 'ene' and 'yne', then both of them are tried to be accommodated in the nomenclature as 'enyne' (grammar). The 'yne' gets priority over 'ene' and comes after it instead of 'ynene' as suffix. If it is not possible to include both 'ene' and 'yne' in the suffix, then 'ene' is transferred to the prefix as 'enyl' substiuents. It should be noted that the lowest locant is to be given to the 'yne' whenever possible. But 'Rule of lowest number' is to be remomebered.

Examples

[i] 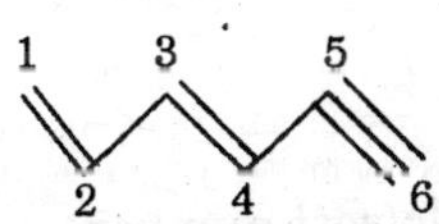

[A]

Sum of locants = 9
(Hexa-1,3-dien-5-yne
starting locant 1)

and

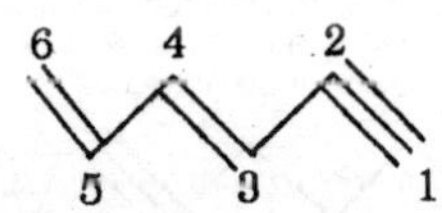

[B]

Sum of locants = 9
(Hexa-3,5-dien-1-yne
starting locant 3)

For [A] and [B] same sum, so smallest number will be the first locant. [A] starts with 1 and [B] with 3. Hence [A] is the correct name.

[ii]  or 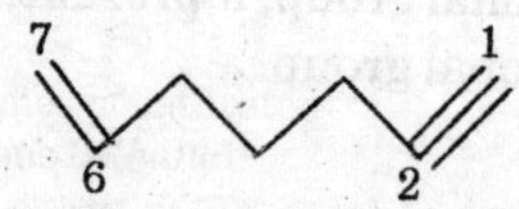

[C]

Hept ~~a~~-1-en ~~e~~-6-yne

= Hept-1-en-6-yne (1+6=7)

[D]

Hept ~~a~~-6-en ~~e~~-1-yne

= Hept-6-en-1-yne (6+1=7)

The numbering in [C] is correct as it starts with lower number 1 instead of 6 in [D]. Priority yne > priority ene, so yne becomes the final (i.e., end) suffix. The name of [D] should not be hept-1-yn-6-ene as it brings the ene as the final suffix.

[iii] 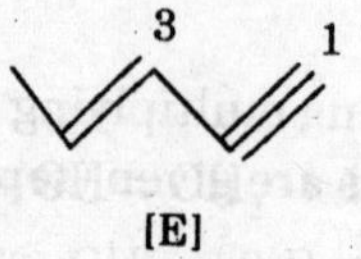 or 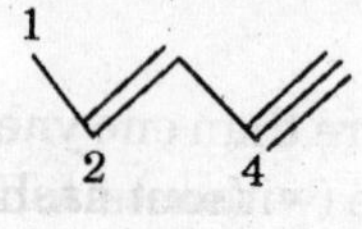

[E]

Sum of locants = (1+3)

Correct numbering

(Pent-3-enyne)

yne–suffix

[F]

Sum of locants (2+4)

Wrong numbering

(Pent-2-en-4-yne)

Here 'Rule of lowest number' is not applicable as the sums of the locants in [E] and [F] are not equal. Hence the 'Rule of lowest sum' is valid.

[iv]  = Correct nomenclature, Pent – 1 – en – 4 – yne (1+4=5, yne end, 4s) / Wrong nomenclature, Pent – 4 – en – 1 – yne (4+1=5, yne end)

[v] 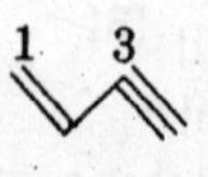 = Correct nomenclature (not But-1-en-3-yne), butenyne (no locant). / Wrong nomenclature, But-3-en-1-yne

[vi]  = Hepta-1, 3, 5-triyne (locants necessary)

[vii] 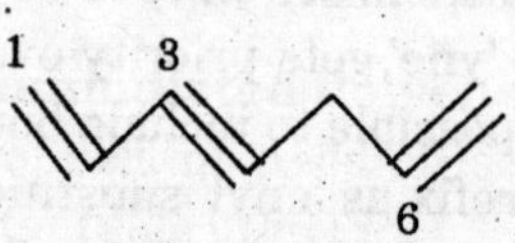 = Hepta-1,3,6-triyne (*not* hepta-1,4,6-triyne)

In hexatriyne locants are unnecessary but in heptatriyne they are needed.

[viii]  = Hept-1-en-5-yne (Rule of lowest sum).

[ix] 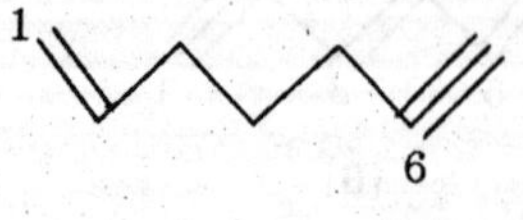 = Hept-1-en-6-yne (Rule of lowest number)

[x] 7 5 1 6  = 6-Methylhept-5-en-1-yne

or 2-Methylhept-2-en-6-yne (Correct, Rule of lowest sum)

[xi] 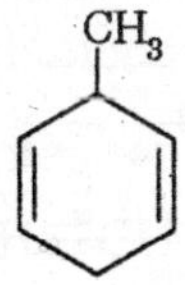 = **1-Methylcyclohexa-2,5-diene (proposed; '4s')**

3-Methylcyclohexa-1,4-diene (wrong)

[d] Name of the important radicals derived from unsaturated compounds

[i] $CH_2{=}CH_2$ (ethene) $\xrightarrow{-H^\bullet}$ $CH_2{=}CH{-}$ [ethenyl; (–e+yl); common name vinyl group]

[ii] $HC{\equiv}CH$ (ethyne) $\xrightarrow{-H^\bullet}$ $CH{\equiv}C{-}$ [ethynyl; (–e+yl) group]

[iii] $CH_2{=}CH{-}CH_3$ (propene) $\xrightarrow{do.}$ $\overset{3}{C}H_2{=}\overset{2}{C}H{-}\overset{1}{C}H_2{-}$ (2-propenyl, C=C at C_2)

[iv] $CH_3{-}CH{=}CH_2$ (propene) $\xrightarrow{do}$ $\overset{3}{C}H_3{-}\overset{2}{C}H{=}\overset{1}{C}H{-}$ (1-propenyl, C=C at C_1)

[v] $HC{\equiv}C{-}CH_3$ (propyne) $\xrightarrow{do.}$ $HC{\equiv}C{-}CH_2{-}$ (2-propynyl, C≡C at C_2)

[vi] $CH_3{-}C{\equiv}CH$ (propyne) $\xrightarrow{do}$ $CH_3{-}\overset{2}{C}{\equiv}\overset{1}{C}{-}$ (1-propynyl, C≡C at C_1)

Examples

[i] 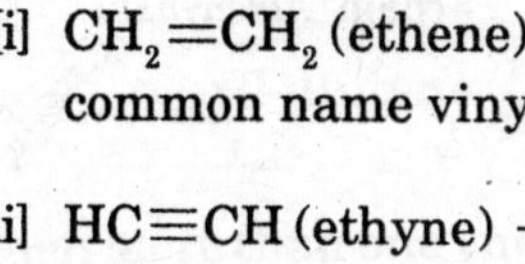 **= 3-Ethenylhepta-4,6-dien–1–yne.**

The chain containing C≡C is to be taken instead of three double bonds in the chain. So ethynyl is not the prefix (i.e., substituent) on the carbon chain.

[ii] 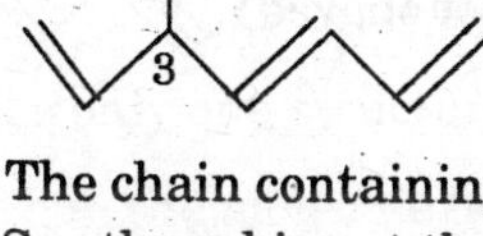 **= 4-Ethenylhept-1-en-5-yne (more acceptable), *or***

4-Vinylhept-1-en-5-yne.

General nomenclature

[a] Alkane (C_nH_{2n+2}) $\xrightarrow{[-2H]}$ Alkene (C_nH_{2n}) $\xrightarrow{[-2H]}$ Alkyne (C_nH_{2n-2})

$\begin{bmatrix}-ane\\+yl\end{bmatrix}\downarrow[-H^\bullet]$ Alkyl; (yl) $(C_nH_{2n+1})^\bullet$

$\begin{bmatrix}-e\\+yl\end{bmatrix}\downarrow[-H^\bullet]$ Alkenyl; (enyl) $(C_nH_{2n-1})^\bullet$

$\begin{bmatrix}-e\\+yl\end{bmatrix}\downarrow[-H^\bullet]$ Alkynyl; (ynyl) $(C_nH_{2n-3})^\bullet$

[b] *Root name of the compound* + ane ⟶ IUPAC names of saturated hydrocarbons (*Alkane*)

+ ene ⟶ IUPAC names of olefinic hydrocarbons (*Alkene*)

+ yne ⟶ IUPAC names of hydrocarbons holding ene and yne on the longest chain (*Alkenyne*)

+ yne ⟶ IUPAC names of hydrocarbons holding (C≡C) in the chain (*Alkyne*)

Examples

[i] $CH_3—CH_2—CH=CH_2$ → **1-Butene or But-1-ene (*not* 1-Butanene)**

[ii] $CH_3—C\equiv C—CH_3$ → **2-Butyne or But-2-yne (*not* 2-Butanyne)**

[iii] $CH_2=CH—C\equiv CH$ → **Butenyne (no locant, *not* Butaenyne).**
So no (ane+ene) or (ane+yne) or (ane+ene+yne), but may (ene+yne) as suffix.

[e] **Common errors: The compound taken as example is [A] = Hexa-1,3-dien-5-yne.**

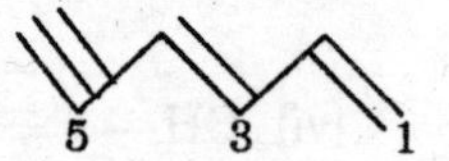

[A]

Errors: [i] **Hexa-1-yne-3,5-diene (e in yne not dropped; yne came before ene, also 'a' in hexa not dropped).**

[ii] **Hexa-1,3-diene-5-yne (the last e of diene not dropped)**

[iii] **Hexa-5-yne-1,3-diene (a in hexa not dropped, yne came before ene)**

[iv] **Hex-1,3-dien-5-yne (a in hexa dropped)**

[v] **Hexa-3,5-dien-1-yne (Rule of lowest number not obeyed)**

[vi] **Hexa-1 3-dien-5-yne (comma dropped)**

[vii] **Hexa 1,3-dien-5-yne (dash dropped)**

[viii] **Hexa-1,3-dien-5yne (dash dropped).**

[f] **Cummulated Polyenes:** $CH_2=C=C=C=C=CH_2$= Hexapentaene (a in penta is not dropped, no need of locants). The stereochemical notations of alkenes will be given later (vide infra).

14. Nomenclature of Alicyclic Compounds

These compounds are called cyclic compounds and all members of the cycle are carbon atoms. The root names of these compounds are derived from the corresponding open chain compounds with the prefix 'cyclo' added to it without any hyphen. Thus cycloalkanes, cycloalkenes and cycloalkynes have carbon and hydrogen in the composition as C_nH_{2n}, C_nH_{2n-2} and C_nH_{2n-4}, respectively. It is to be noted that cycloalkanes are isomeric with alkenes both having the degree of unsaturation one. The nomenclature of the clyclic compounds is similar to that of open chain compounds.

Examples

Butane, C_4H_{10} → Cyclobutane C_4H_8 → But-2-ene C_4H_8 → Cyclobutene C_4H_6 (no locant)

The characteristic of the cyclic compound is that when started from a particular member and progressed continuously the starting point (or member) will be reached again. When the functional groups are present then these groups will be either suffix or prefix with locants and all the rules applicable to the open-chain compounds will be followed.

Examples

[i]

Cyclopropane
Saturated (C_3H_6)-ane

Cyclopropene
Unsat. (C_3H_4)-ene
(no locant)

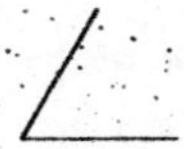

Propane
Sat.(C_3H_8)-ane

[ii]

 ;

Cyclopentane*
(C_5H_{10}) Sat. (ane)
* = [a in penta dropped]

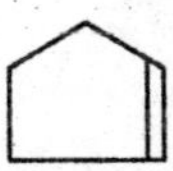 ;

Cyclopentene*
C_5H_8 (Unsat. ene)
(no locant)

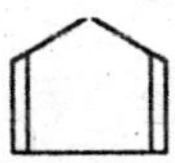 ;

Cyclopentadiene*
C_5H_6 (di.ene)
(no locant) [a stays]

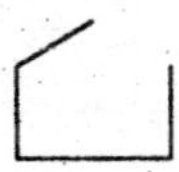 ;

Pentane
Sat. (C_5H_{12}) ane

[iii]

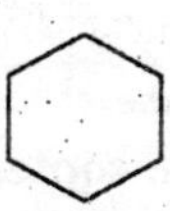

Cyclohexane*
C_6H_{12}
Sat.(ane)
*(a in hexa dropped)

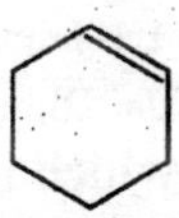

Cyclohexene*
C_6H_{10}
(Unsat.ene)
*(a in hexa dropped)
(no locant)

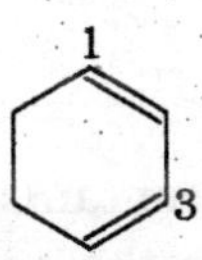

Cyclohexa-1,3-diene
C_6H_8
(di.ene)
(locants needed)
(a stays)

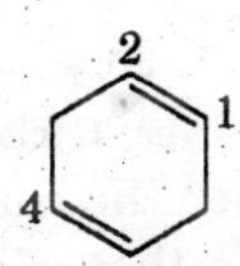

Cyclohexa-1,4-diene
C_6H_8
(locants needed)

[iv]

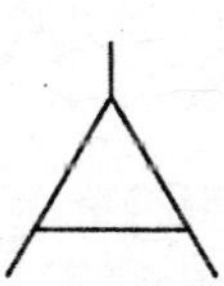

[Substituted Cyclo = *]

1,2,3-Trimethylcyclopropane* (Single word, no dash between Tri, methyl, cyclo and propane).

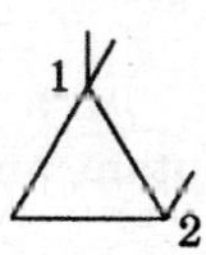

1,1,2-Trimethylcyclopropane*

[v] 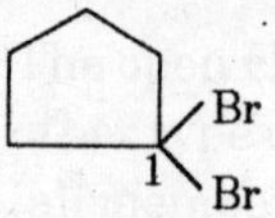 1,1-Dibromocyclopentane*

[vi] 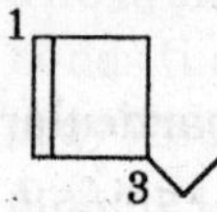 3-Ethylcyclobutene*, (the locant of 'ene' is unnecessary)

[vii] 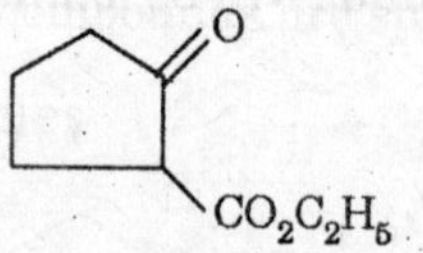 1-Carbethyoxy-2-ketocyclopentane* (better name)

or 2-Ethoxycarbonylcyclopentanone

or 2-Carbethoxycyclopentan-1-one*

[viii] [A] 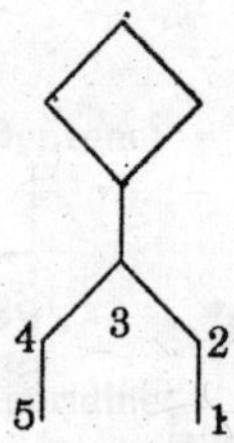 [B] 

3-Cyclobutylpentane; (pr. Cyclo.butyl.pentane)

1,1–Dimethylcyclobutane* (pr. Di.methyl.cyclo.butane)

When the ring (i.e., cycle) contains a branching which has the carbon content larger than that of the ring, then the name of the compound will be as the derivative of alkane (e.g., A), but if the cycle contains greater number of carbon than the side chain, then the name will be derived from cycloalkane (e.g., B).

[**Comments**: The authors feel that 3-cyclobutylpentane (viiiA) but as (viii C), may be named as 3,3-trimethylenopentane or 1, 1-diethylcyclobutane].

[C] 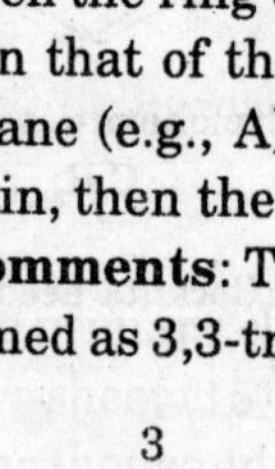

[ix] 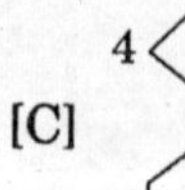 = 1,2-Dimethylcyclobutane; = 1,3-Dimethylcyclobutane and

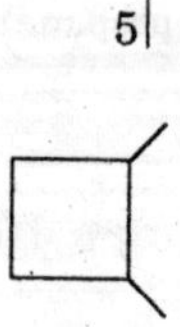 = 1,1-Dimethylcyclobutane (locants necessary).

[x] 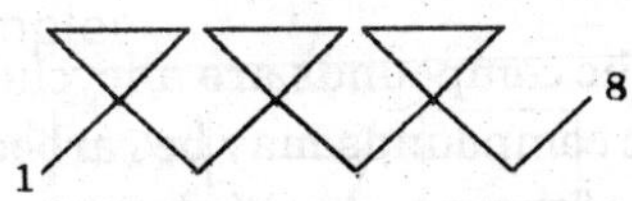 = **2,4,6-Triethanooctane;**

[xi] 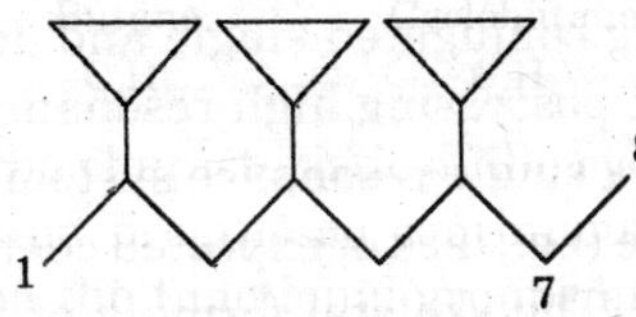 = **2,4,6-Tricyclopropyloctane**

[xiA] 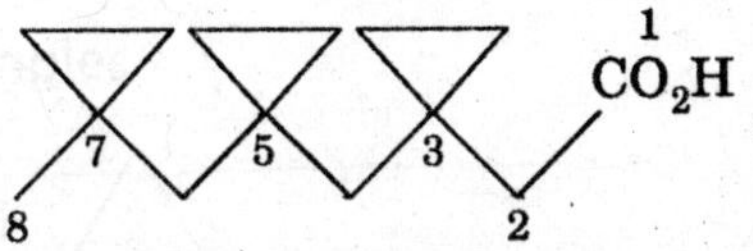 = **3,5,7-Triethanooctanoic acid (CO_2H has the C_1)**

[xii] 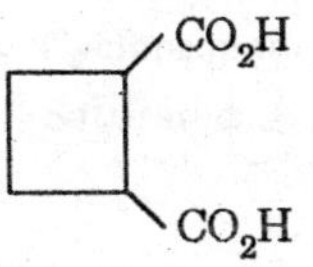 = **Cyclobutane-1,2-dicarboxylic acid**

or **1,2-Dicarboxycyclobutane**

[xiii] 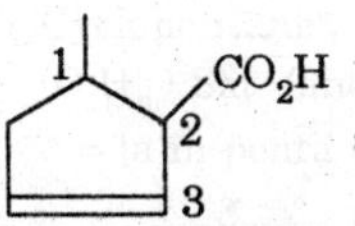 = **2-Carboxy-1-methylcyclopent-3-ene**

(alphabetic order, locants and rule of lowest sum and other formalities applicable to aliphatic compounds have been observed)

[xiv] 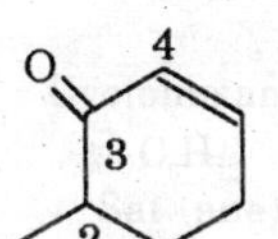 = **1-Ethyl-3-keto-2-methylcyclohex-4-ene (vide xiii above).**

[xv]

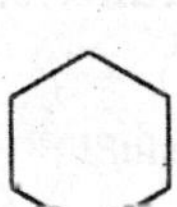

A monocyclic compound

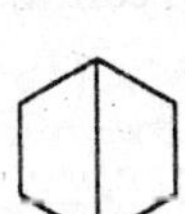

A bicylic compound

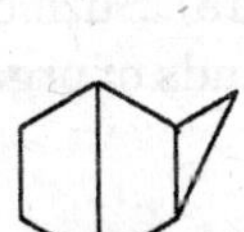

A tricyclic compound

[xvi] 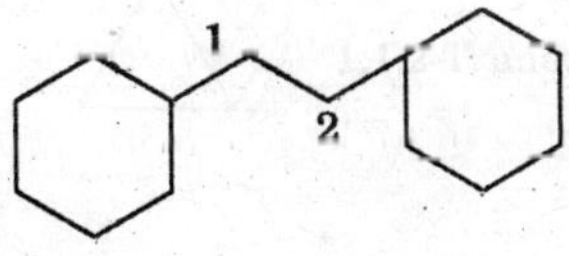

1,2-Dicyclohexylethane

(Presence of more than one ring will make the name as a derivative of alkane).

The open chain compounds are aliphatic and the cyclic compounds are alicyclic. Another type of cyclic compounds is aromatic. The cyclic compounds may be carbocyclic (all members are carbon) or heterocyclic (at least one of the members is heteroatom). The aromatic compounds are cyclic molecules having conjugated single and double bonds (i.e., alternately single and double bonds) and possessing high resonance energy where all the members of the cycle take part in a single resonance system thus making the aromatic molecules extraordinarily stable (i.e., lack of expected chemical reactions characteristics of the olefinic compounds). The skeletal formula of these compounds are shown below.

Examples

[i] 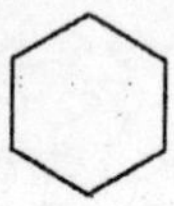 Cyclohexane; 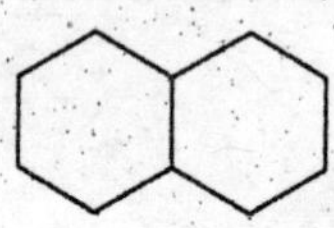 Decahydronaphthalene (Alicyclic)

(also called decalin)

[ii]  Benzene; 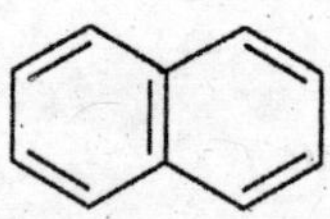 Naphthalene (Aromatic)

[iii] 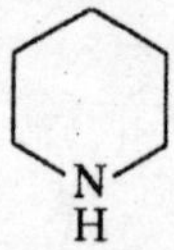 Piperidine; 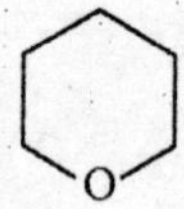, , etc.

(Saturated heterocyclic, i.e., aliphatic heterocycles).

[iv]  Pyridine; 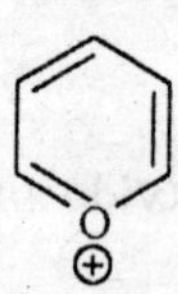, 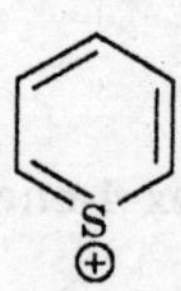, etc. (Aromatic heterocycles)

The primary classification of the organic compounds has been shown in the following flow-sheet (15). Such compounds when connected with various functional groups deliver thousands of organic compounds.

15. Flowsheet of Classification of Organic Compounds

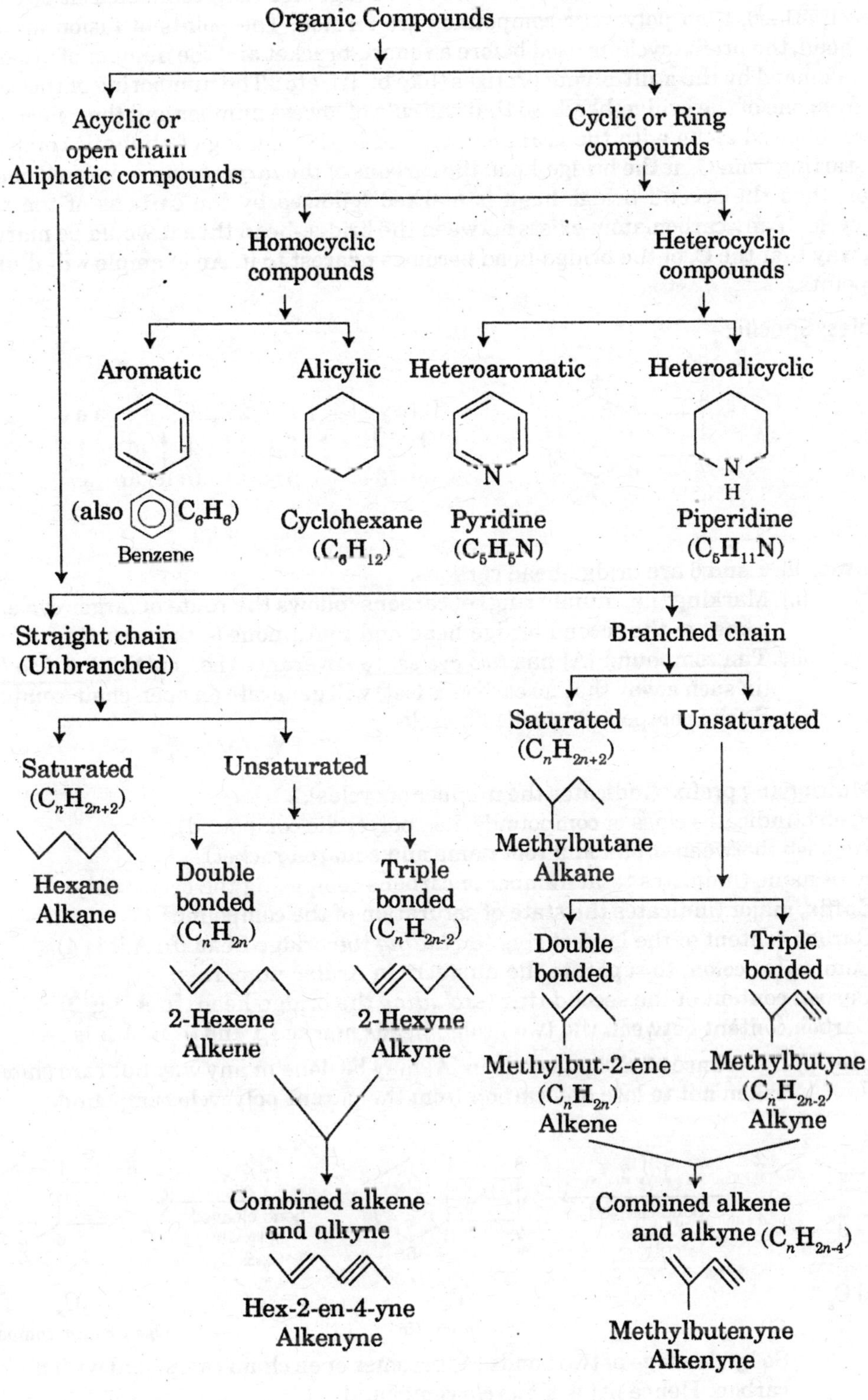

16. Nomenclature of Polycyclic Compounds

When more than one cyclic compounds are fused together (i.e., connected at two points with each other), then polycyclic compounds are formed. The points of fusion are called bridge-head, the prefix 'cyclo' is used before a square-bracket and the number of cycles thus fused is denoted by the multiplying prefixes (like bi, tri, etc). The numbering of the carbons starts from one of the bridge-heads so that the rule of lowest number and the rule of lowest sum are followed along with the grammar applied in the naming of aliphatic compounds. Thus, starting from C_1 at the bridge-head the carbons of the *largest ring* is marked one after another, then the second bridge-head is marked followed by the carbons of the second larger cycle. If any carbon atom exists between the bridge-head then it would be marked in such a way that the C_1 of the bridge-head becomes nearest to it. An example will illustrate these points.

Examples: Specific

[I]

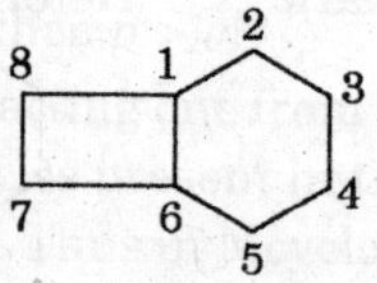

[A]

Bicyclo [4 , 2 , 0] octane
(a) (b) (c) (f) (g) (h) (i) (c) (d) (e)

Name of [A]

Notes: [i] 1 and 6 are bridge-head carbons.
[ii] Marking (i.e., numbering) of carbons follows the route of larger cycle, then comes to the second bridge-head and continuous to the next ring.
[iii] The compound [A] has *two cycles*, so severance (i.e., cutting) of *two bonds* (in such a way that no carbon is lost) will generate an open chain compound. So the compound is called bicyclo.

Legends:
[a] Multiplying prefix: (indicates the number of cycles).
[b] Prefix (indicates class of compounds, i.e., polycyclic compound).
[c] No dash (between prefix and root name and square bracket).
[d] Root name (indicates total number of carbons composing the cycles only).
[e] Suffix, major (indicates the state of saturation of the compound).
[f] Carbon content of the largest ring (*excluding* the bridge-head. In A it is 4).
[g] Comma (or colon, to separate the numbers in Arabic numerals).
[h] Carbon content of the second ring (*excluding* the bridge-head. In A it is 2).
[i] Carbon content between the two bridge-heads marked 1 and 6 (In A it is 0).

Note: [i] Severance of carbon bonds in [A] may be done in any way but care should be taken not to lose any carbon from the parent polycyclo compound.

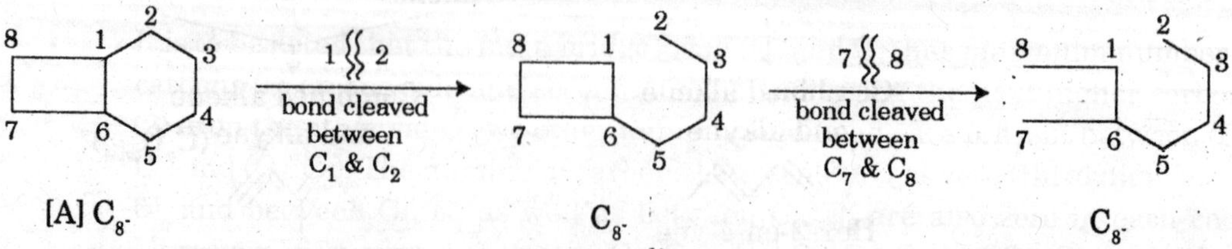

[A] C_8-　　　C_8- *monocyclic*　　　C_8- *Open chain* compound

So by cleavage of two bonds [A] becomes open chain compound with *no loss* of carbon. Hence [A] is a *bicyclo* compound.

[II] Another example

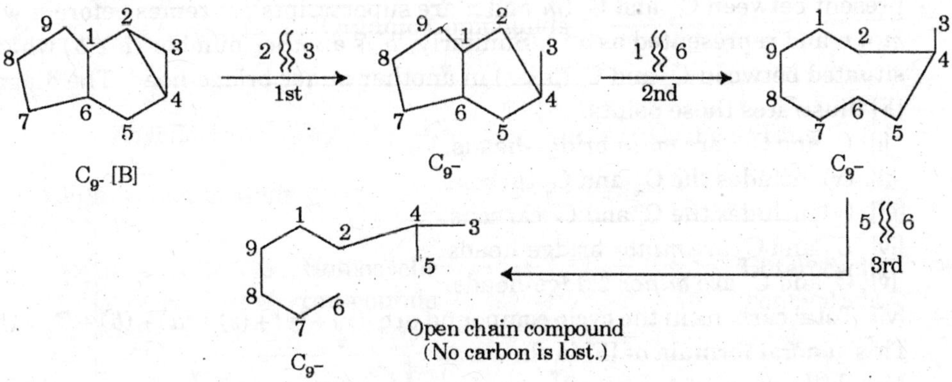

So [B] is a tricyclo compound.

Thus it becomes very easy to determine the multiplying prefixes of the compounds [C], [D] and [E] as *bi*cyclo, *tetra*cyclo and *penta*cyclo, respectively.

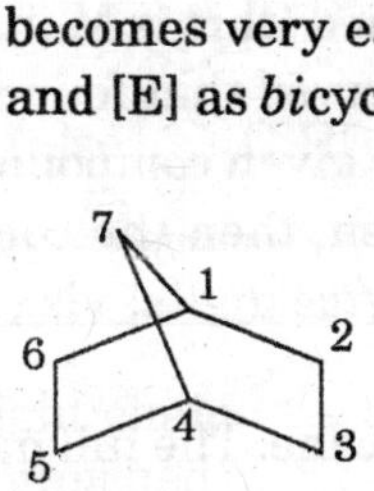

Bicyclo [C]

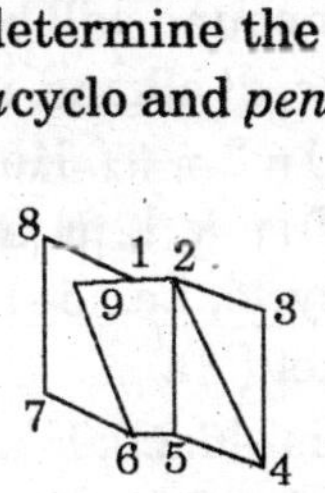

Tetracyclo [D]

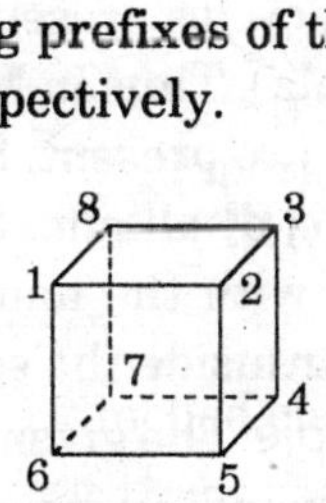

Cubane

Pentacyclo [E]

[III] General example: This should be very useful in understanding the method of nomenclature of the polycyclic compounds. The following example is taken: Mcyclo [$x : y : z : a^{m,n} : b^{p,r}$] alkane.

[i] M = Multiplying prefix; will show the number of cycles in the compound.

[ii] No dash before and after the square bracket [ ].

[iii] x is the number of carbon atoms ($x \neq 0$) on either side of the bridge-head, excluding the bridge-head carbons (i.e., the carbons on the ring juncture). x may be the largest or equal to y but never smaller than y. Thus $x \geq y$.

[iv] y is the number of carbons on the opposite side of x and excludes the bridge-head carbon just like x; y is the second large number ($y \geq z$).

[v] z is the number of carbon(s) between the two main bridge-heads (i.e., between C_1 and C_{x+2}). z may be zero and never larger than x or y.

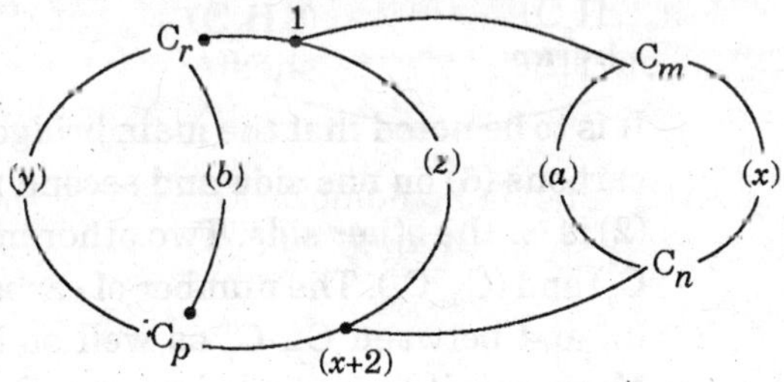

[F]

[vi] In addition to the principal (i.e., main) bridge-head (holding z in between), there may exist *secondary* bridge-head(s). The 'a' represents the number of carbons present between C_m and C_n (m and n are superscripts), m comes before n when $m < n$ and represented as $a^{m,n}$. Similarly, 'b' is another number ($a \geq b$) which is situated between C_p and C_r ($p < r$) in another minor bridge-head. The diagram [F] illustrates these points.

[i] C_1 and C_{x+2} are *main* bridge-heads.
[ii] (x) includes the C_m and C_n carbons.
[iii] (y) includes the C_r and C_p carbons.
[iv] C_m and C_n are *minor* bridge-heads.
[v] C_r and C_p are *minor* bridge-heads.
[vi] Total carbons in the cyclo compound are $[(x)+(y)+(z)+(a)+(b)+C_1+C_{x+2}]$.

This general formula of [F] is shown as Mcyclo [x, y, z, $a^{m,n}$, $b^{p,r}$] alkane. Here, M = Tetra, and ($x \geq y \geq z \geq a \geq b$, if $a = b$, then $p > m$).

[vii] By subtracting one from the digits present inside the third bracket [ ] the number of cycles present in the compound will be obtained (this is M in the general formula). Thus in Mcyclo [a, b, c, d] alkane, it is observed that four digits, a, b, c and d are present. So M = (4–1) = 3 = *tri*. Hence this given compound is tricyclo [a, b, c, d] alkane. Similarly, if [j, k, l, m, n] is given, then the compound will begin with the name as tetracyclo, i.e., 5–1 (since *five* digits, viz., j, k, l, m, n appear inside the square bracket [ ]).

Example: The given compound is –[6:2:0:0^{2,7} : 0^{3,6}] decane. The full name and the corresponding structural formula of the given compound can be determined as follows:

Since there are five digits, viz., 6,2,0,0,0, inside the [ ] the compound will be (5–1) = tetracyclo [6, 2, 0, 0^{2,7}, 0^{3,6}] decane (same as 6:2:0:0^{2,7}:0^{3,6}). The drawing will be as follows,

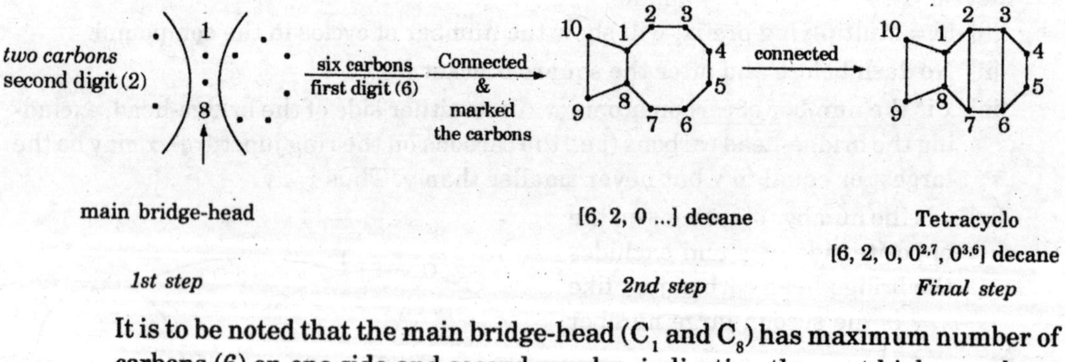

It is to be noted that the main bridge-head (C_1 and C_8) has maximum number of carbons (6) on one side and second number indicating the next higher carbons (2) is on the other side. Two other minor bridge-heads are present between (C_2, C_7) and (C_3, C_6). The number of carbons between C_1, C_8 is zero (third digit shows 0), and between C_2, C_7 as well as between C_3, C_6 are also zero in each case. Moreover, if four bonds between C_9, C_{10}; C_1, C_8; C_2, C_7 and C_3, C_6 are broken, then the open chain compound containing 10 carbon atoms is obtained (i.e., no carbon is lost during the bond breaking). Hence the molecule is tetracyclo.

[viii] The total sum of numbers inside the [ ] and two are added to the total sum. This final total will determine the number for carbons constructing the polycyclic compound. Thus, if the general formula is given as,–[x+y+z+a+b]–, then the total number of carbon atoms constructing the polyclic ring will be (x+y+z+a+b+2)= root name of the compound (excluding carbon-containing functional groups), with the major suffix like ane, ene or yne (also enyne) as the nomenclature demands. Minor suffixes, if present, will be added after the major suffix.

Examples

[a] The notation of the compound is $-[5:3:0, 0^{2,5}:0^{8,10}]-$. The complete structure and nomenclature of the given notation are to be furnished.

Procedure: The total number of carbons constructing the polycycle is (5+3+0+0+0+2)=10=root name is deca. Assuming that the compound is saturated the major suffix will be 'ane'. So the compound is decane (dec~~a~~+ane).

There are five digits inside the [ ], viz., 5, 3, 0, 0, 0; hence number of cycles present will be 5–1 (vide vii) = 4 = tetracyclo. So the complete name of the compound is 'Tetracyclo $[5:3:0, 0^{2,5}, 0^{8,10}]$ decane.' [A]

The structural formula will be as follows.

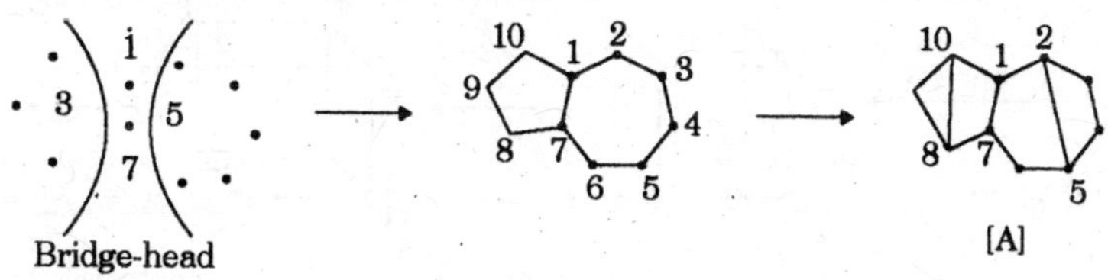

[A]

Since 2 < 8, so $0^{2,5}$ comes before $0^{8,10}$.

[b] 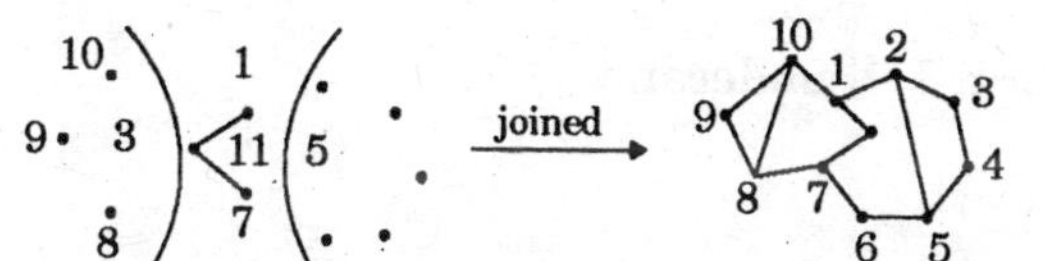

One C between 1 and 7 (3rd digit)
No C between 2 and 5 (4th digit)
No C between 8 and 10 (5th digit)

[5:3:1 ....]
1st step

[B]
Final Step

$[5, 3, 1, 0^{2,5}, 0^{8,10}]$—is given notation of the compound. Total number of carbons = (5+3+1+0+0+2) = 11 = undecane (if saturated). Again, total number of digits inside [ ] = 5. So cycles are (5 – 1) = 4 = tetracyclo. So the nomenclature is 'Tetracyclo' $[5:3:1:0^{2,5}:0^{8,10}]$ undecane = [B]. The construction of the structure is done as shown above.

[c] 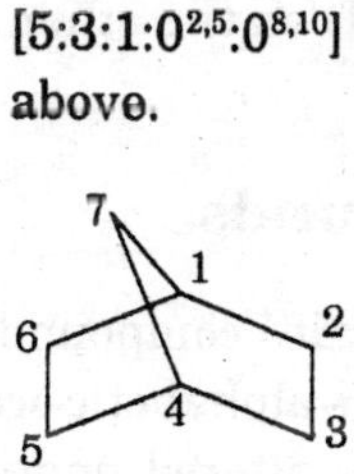 = Bicyclo [2:2:1] heptane.

[d] 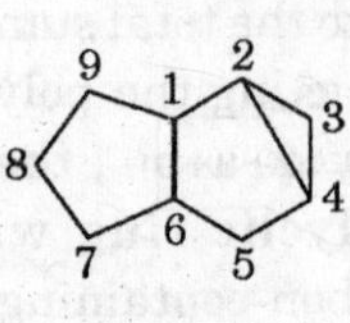 = Tricyclo [4,3,0,$0^{2,4}$] nonane.

(Four digits [ ], so tricyclo; 4+3+2=9=nonane.)

[e] 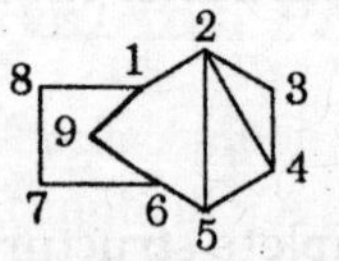 = Tetracyclo [4:2:1:$0^{2,4}$; $0^{2,5}$] nonane [$0^{2,4}$, comes before $0^{2,5}$, Since 4 < 5]

[f] The structure and nomenclature of,–[4, 2, 0, $0^{2,5}$, $0^{3,8}$, $0^{4,7}$]— is to be given.
The name of the compound should be (6 – 1) = 5 = Pentacyclo.
The number of carbon atoms should be [4+2+0+0+0+0+2] = 8 = Octa = root name.
If the molecule be saturated, the major suffix is 'ane', so ending is octane. Therefore the nomenclature is pentacyclo [4:2:0:$0^{2,5}$, $0^{3,8}$, $0^{4,7}$] octane.
Drawing the structure in stepwise manner,

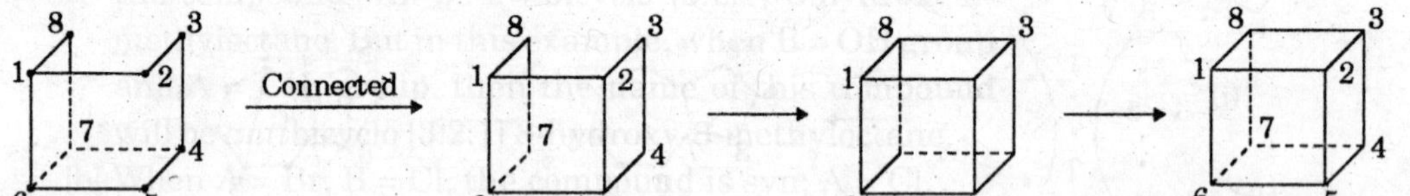

1,6-Bridge-head —[4, 2, 0, ...] octane 1st step | —[4, 2, 0, $0^{2,5}$, ...] octane 2nd step | 3rd step | pentacyclo [4, 2, 0, $0^{2,5}$, $0^{3,8}$, $0^{4,7}$] —octane = Cubane

[g] –[5, 2, 1, $1^{3,5}$]–. This is 'Tricyclo [5, 2, 1, $1^{3,5}$] undecane.'

[h] 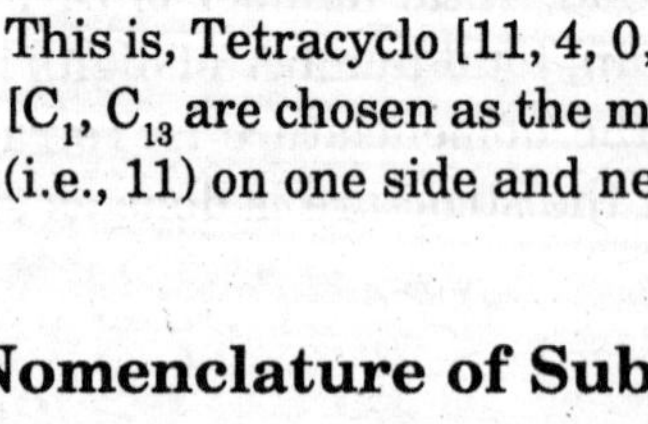

This is, Tetracyclo [11, 4, 0, $0^{2,10}$, $0^{5,9}$] heptadecane.
[C_1, C_{13} are chosen as the main bridge-head carbons to hold maximum carbon atoms (i.e., 11) on one side and nearest to the maximum carbon (i.e., 4) on the other side].

17. Nomenclature of Substituted Polycyclic Compounds

The well-known rules for the nomenclatures of acyclic (i.e., open chain) compounds are followed including the rule of lowest sum, rule of lowest number and the alphabetic order of the substituents but the digits inside the square bracket must not be altered under any

circumstances. Thus the main bridge-head carbon must be marked as C_1 and the locants of the substituents should be given the lowest possible number. The following examples illustrate these points.

Examples

[i] 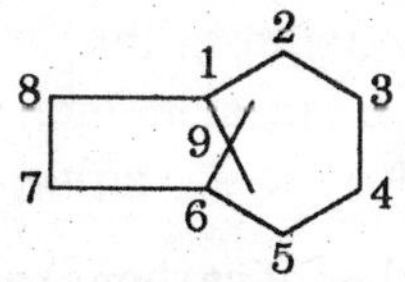

This marking of the carbon atoms *must* be followed. Hence the name is bicyclo [4, 2, 1] 9,9-dimethylnonane. (*not* 2,2-dimethylnonane).

[ii] 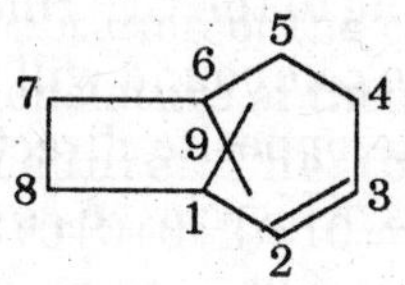

This is bicyclo [4:2:1] 9,9-dimethylnon-2-ene. (*not* 9,9-dimethylnon [4:2:1]-4-ene)

[iii] Drawing the structural formula for bicyclo [4, 2, 1] 9-methyl-3-oxonon-7-yne (also called bicyclo [4, 2, 1] 3-keto-9-methylnon-7-yne)

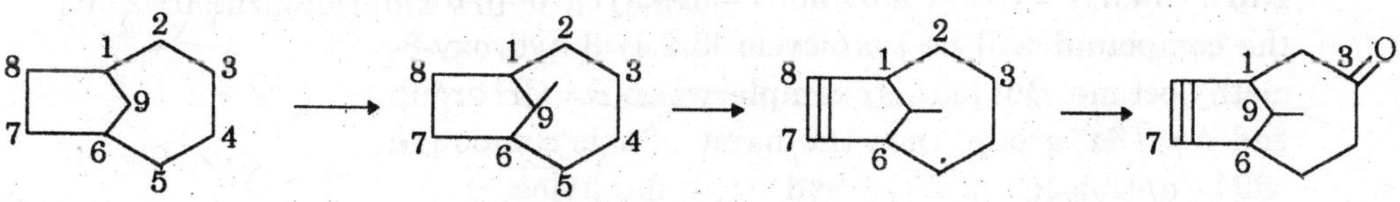

Frame of bicyclo compopund with marking of the carbons (1st step) — **(2nd step) -9-methyl** — **(3rd step) -7-yne** — **(Final step, the given compound)**

[iv] Bicyclo [4, 2, 1] 8-methylnon-2-en-9-one [A] is

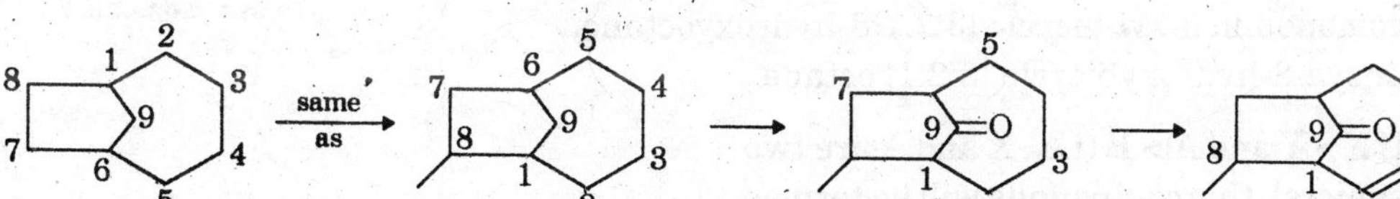

(8-methyl) — **(8-methyl-9-one)** — **[A]**

[v] 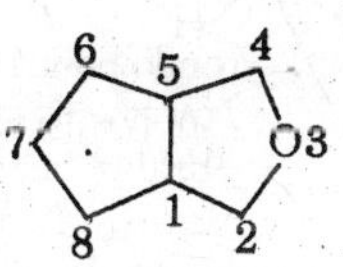

This is bicyclo [3, 3, 0] 3-oxaoctane (*not* oxooctane or oxoctane).

[vi] 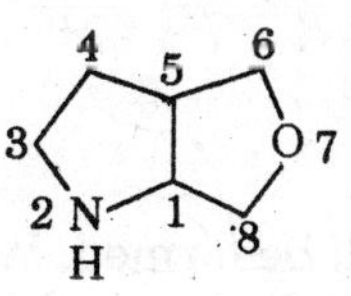

This is bicyclo [3, 3, 0] 7-oxa-2-azaoctane. The priority oxa > aza, so oxa comes before aza (alphabatic order not followed)

[vii] 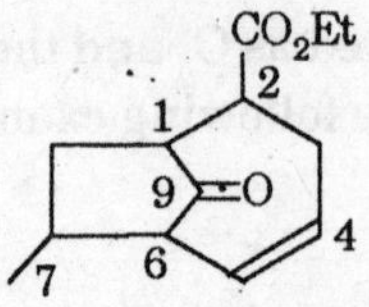

This is bicyclo [4, 2, 1] 2-carbethoxy-9-keto-7-methylnon-4-ene (*not* bicyclo [4, 2, 1]-5-carbethoxy-9-keto-8-methylnon-2-ene)

18. Syn/Anti Prefixes of Polycyclic Compounds

In a bicyclo compound like [X, Y, Z], let X > Y > Z (based on number of carbons) and Z holds substituents like A and B (A has greater priority than B according to the CIP rules). Now if two seniors X and A are on the same side of the bridge-head the compound will be termed *syn*-bicyclo. On the contrary when A (A > B) and X are on the opposite direction of the bridge-head, then the stereochemical notation will be *anti* as prefix of bicyclo (i.e., *anti*bicyclo).

Examples

[i] [a] Here, X > Y (X holds 3 carbons > Y holds 2 carbons) and by CIP rules A > B (priority), the compound is *syn*bicyclo. Thus, when A = OH group and B = methyl group, then the compound will be *syn*bicyclo [3:2:1] 8-hydroxy-8-methyloctane. But in this example, when B = OH group and A = CH_3 group, then the name of this compound will be *anti*bicyclo [3:2:1] 8-hydroxy-8-methyloctane.

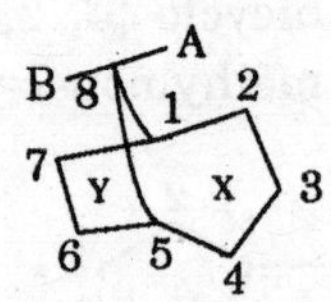

(i)

[b] When A = Br, B = Cl, the compound is syn; A = Cl, B = OH, give *syn*-compound,
but A = Br, B = I, the compound is *anti*; A = CCl_3, B = CH_2Br give *anti*–[vide C.I.P. rules 52 (b)].

[ii] Here X > Y, and OH > H, so the name of the given comopound is *syn*bicyclo [3:2:1]8-hydroxyoctane. or *syn*-8-hydroxybicyclo [3:2:1] octane.

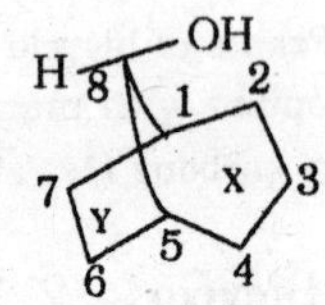

(*syn* isomer ii)

[iii] If X > Y and A > B (i.e., X and A are two seniors), this compounds will be termed *anti*. Thus in the given compound, if A = Cl and B = OH (here A > B by CIP rules), the compound is *anti*-bicyclo [3:2:1] 8-chloro-8-hydroxyoctane. When the positions of OH and Cl are exchanged then the compound will be converted into the *syn*bicyclo.

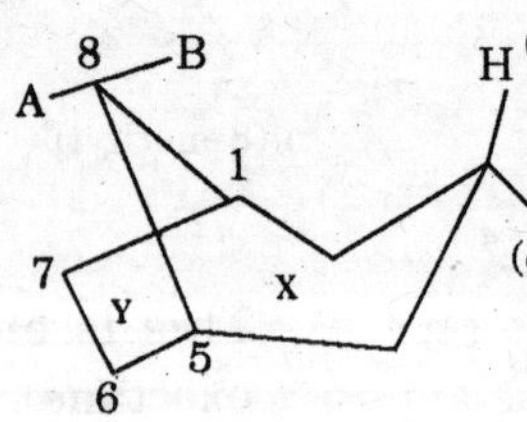

19. Nomenclature of Spiro Compounds

When a single carbon holds two rings, then a spiro compound will be formed. When these two rings are of different sizes (i.e., the number of carbons making the rings are different), then the numbering of the carbons of the compounds starts from the nearest position of the central carbon (i.e., the carbon holding these two rings to make the 'spiro' compound), proceeds through each member of the smaller ring, marks the central carbon, enters the larger ring to give the lowest locant to the substituents and comes back to the central carbon. The rule of lowest sum and rule of lowest number as well as the alphabetic order are obeyed. The IUPAC name starts with the prefix 'spiro' followed by [ ] and suffixed in the method of open-chain-compound nomenclature. The following examples will explain all these points.

[A]

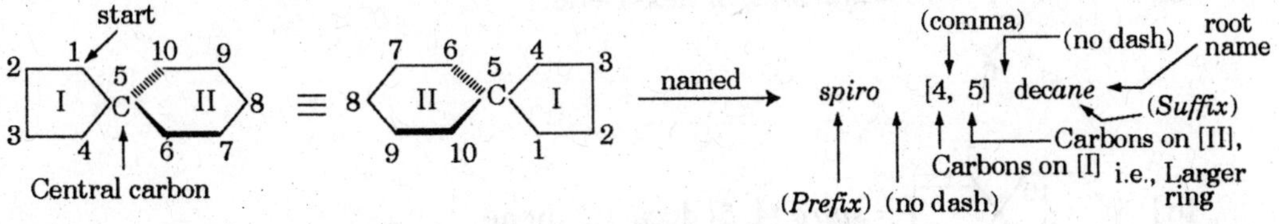

The two rings are at right angles to each other (orthogonal).

[i] [I] = 5-membered ring
[ii] [II] = 6-membered ring
[iii] 5 < 6, so making starts on [I].
[iv] Central carbon is 5.
[v] Marking on [I] starts from nearest carbon of 5.

[i] [4+5+1]=10=*Decane* ← suffix
↑
root name

[ii] Root name = 4 (members of small ring) + 5 (members of large ring) + 1 (central carbon) = 10

[B] General formula

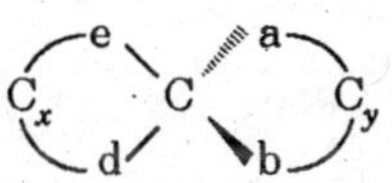

= *Spiro* [(C_y + a + b), (C_x + e + d)] alkane

(no dash) Smaller ring; Comma; Larger ring; (no dash); Suffix with root name; root name

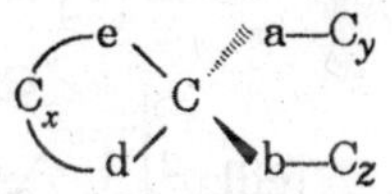

(Cyclic, not spiro compound)

[i] C is central carbon.
[ii] The plane holding a–C_y–b is at right angle to the plane holding d–C_x–e.
[iii] If Y < X, then the notation will be [Y, X] and the name will be spiro [Y, X] alkane. But when Y > X then name will be spiro [X, Y] alkane.
[iv] Y = (C_y + a + b); X = (C_x + d + e) and root name = [X + Y + 1]; (1 is for central carbon).

[C] Specific examples

[i] 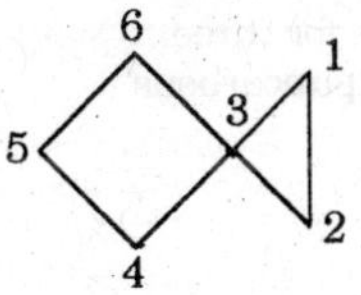 = *spiro* [2, 3] hexane.

[ii] 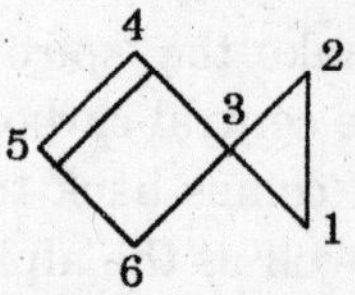 = *spiro* [2, 3] hex-4-ene (Rule of lowest number, 4<5)

(*not spiro* [2,3] hex-5-ene or *spiro* [3, 2] hex-4-ene or *spiro*-[2,3]-hex-4-ene or [2, 3] hex-4-ene).

[iii] 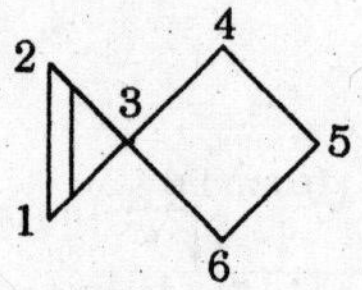 = *spiro* [2, 3] hex-1-ene.

[iv] 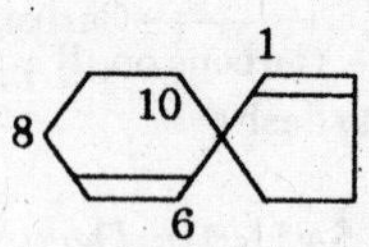 = *spiro* [4, 5] deca-1,6-diene.

[v] 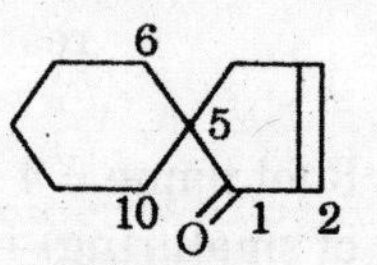 = *spiro* [4, 5] dec-2-en-1-one

[vi] 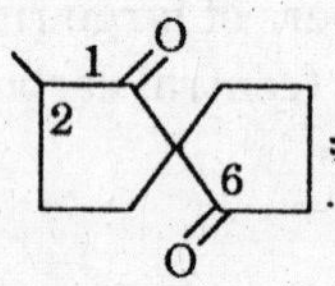 = *spiro* [4, 4] 2-methylnonane-1, 6-dione.

[vii] 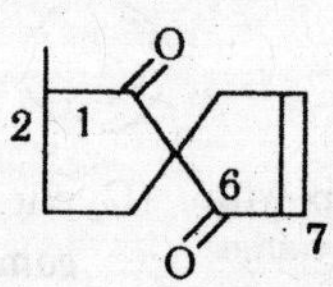 = *spiro* [4, 4] 2-methylnon-7-ene-1, 6-dione.

(Rule of lowest numbers)
= *spiro* [4, 4] 7-methylnon-2-ene-1, 6-dione.

[a] General formula [A]

[i] a through f = number of carbon atoms.

[ii] number of spirocarbons (may be any number) = Multiplying prefix (MP).

Name, Trispiro [a:b:c:d:e:f] (a+b+c+d+e+f+3) alkane

(b < f)

↑ M.P. ↑ Class ↑ No Dash ↑ No Dash ↑ for three spirocarbons

[i] Correct direction → (b < f)

[ii] Wrong direction ········▸ (proposed 4s)

[iii] Total carbons (17 + 3) = 20 = eicosa.

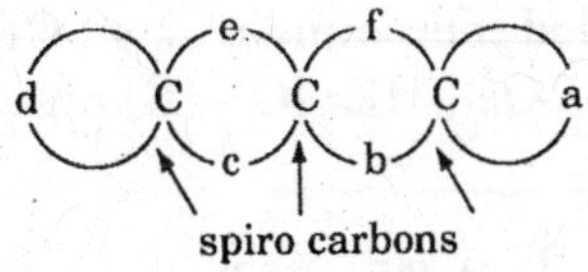

[A]

Correct path is continuous bent arrow not the dotted bent arrow (in [i])

[b] **Examples** (Polyspiro compounds[18,21])

[i] 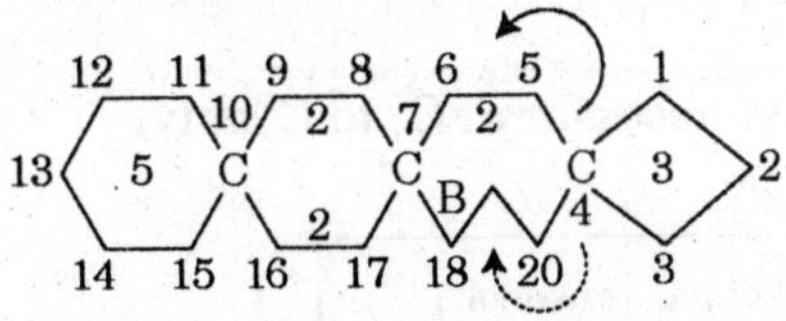

Trispiro [3 : 2 : 2 : 5 : 2 : 3] eicosane (17+3), *not* trispiro [3 : 3 : 2 : 5 : 2 : 2] eicosane.

[ii] 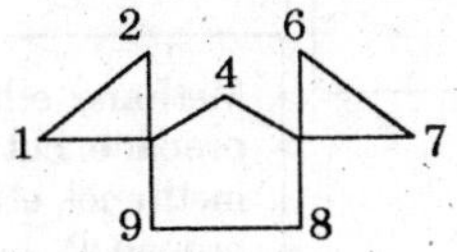 is dispiro [2 : 1 : 2 : 2] nonane.

Note: 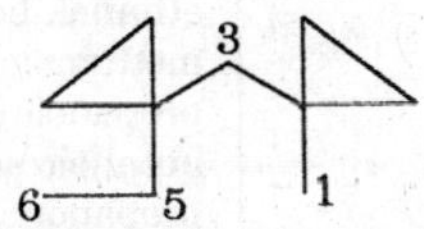 This is not a spiro compound.

This may be called, 2,4-diethanohexane.

20. [A] Common Usages to represent the Suffix and Prefix of the Functional Groups [*p* = prefix, *s* = suffix]

[i] X = halo (*p*), but in binary compounds as 'halide' (e.g., CCl_4 = carbon tetrachloride).

[ii] R = alkyl (*p*); Ar = aryl (*p*).

[iii] H>C=O = al (*s*), formyl (*p*) , carbaldehyde (s), oxo = same as ketone = (*p*).

[iv] C, C>C=O = one (*s*), keto (*p*), oxo (*p*).

[v] —C(OH)=O = carboxy (*p*), oic acid (*s*).

[vi] —C(OR)—O = carbalkoxy (*p*), alkoxycarbonyl (*p*), oate (*s*).

[vii] –OH – hydroxy (*p*), ol (*s*).

[viii] –OR = alkoxy (*p*), oxa (*p*).

[ix] $-NH_2$ = amine (*s*), amino (*p*), azane (*s*).

[x] –CN = cyano, (*p*) nitrile (s), carbonitrile (s).

[xi] $-NO_2$ = nitro (*p*).

[xii] Labelled compounds[21], e.g., $CH_2[^2H]—COOH$ = [2 – $^2H_1$] acetic acid and $CH_3—COO[^2H]$ = [O – ^2H] acetic acid.

[B] Locant(s) 'Yes'; locant(s) 'No'

The following list is useful; [* means locant(s) necessary , C_1 to C_4 only]

Root name
C_1 = metha; C_2 = etha; C_3 = propa; C_4 = buta

only 'ene' in the compound →

Ethene, propene, butene*, ketene, propenol*, butenol*, propenoic acid, butenoic acid*, methyl propenoate, butenone, chloroethene, dichloroethene*, chloropropene*, chlorobutene*, nitrobutene*, aminopropene*, propenenitrile, butenenitrile*, butenamide*

only 'ane' in the compound →

Methane, ethane, propane, butane, methanol, ethanol, propanol*, butanol*, methanal, propanal, ethanal, butanal, methanoic acid, propanoic acid, butanoic acid, propanone, butanone, aminopropane*, aminobutane*, chloropropane*, chlorobutane*, nitropropane*, nitrobutane*, propanenitrile, butanenitrile, propanamide, butanamide, methyl propanoate, methyl butanoate, methyl propane, dimethyl propane, methyl butane, dimethylbutane*, dichloroethane*, pentachloroethane

only 'yne' in the compound →

Ethyne, propyne, butyne*, propynol*, butynol*, propynal, butynal*, butynone, aminopropyne*, chloroethyne, chloropropyne*, chlorobutyne*, nitropropyne*, nitrobutyne*, propynenitrile, butynenitrile*, propynoic acid, butynoic acid*

suffix 'enyne' in the compound →

Butenyne, Halobutenyne*, Nitrobutenyne*, Aminobutenyne*, Arylbutenyne*, Methylbutenyne

The following points are to be noted:

[i] The compounds have any one of 'ane', 'ene', 'yne' and 'enyne'. These are major suffixes, which may be followed by minor suffixes of functional groups.

[ii] The compounds marked with asterisks (*) will exhibit positional isomerism.

[iii] The carbon atoms of –CHO, –CN and –COR (R=OH, OR′, Cl, NH_2) will *always* be marked as C_1. In these compounds the rule of lowest sum and the rule of lowest number are not applicable.

[iv] **If the names do not lead to isomers, then locant(s) should be ignored.**

e.g. [a] **trichloroethene, $Cl_2C{=}CHCl$ → no isomer, no locants (*not* 1,1, 2-trichloroethene)**

[b] **dichloroiodoethene, C_2HCl_2I → isomers, so locants,**

$Cl_2C{=}CHI$ (1,1-Dichloroiodoethene) and $IClC{=}CHCl$ (1, 2-Dichloroiodoethene)

21. Nomenclature of Aromatic Compounds

The common names of a few well-known aromatic compounds are given below. Benzene, C_6H_6, the parent of all carbocyclic aromatic compounds, also represented by Robinson's symbol [A]. The circle inside the regular hexagon indicates six π (*pr. pi.*) electrons.

[i] 

[A] Benzene

[ii] 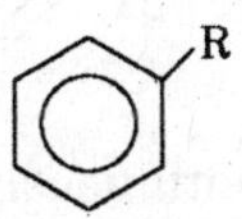

Alkyl benzene

[iii] 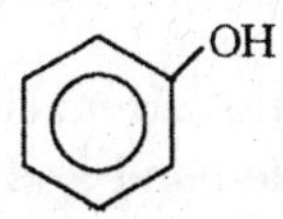

Phenol (Carbolic Acid)

[iv] 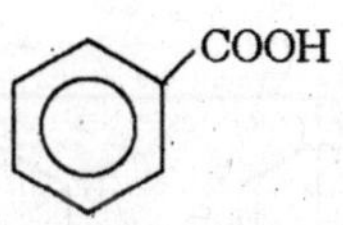

Benzoic Acid

[v] 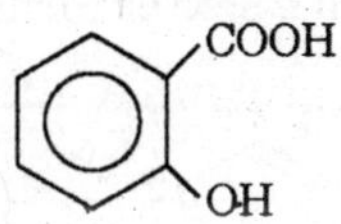

Salicyclic Acid

[vi] 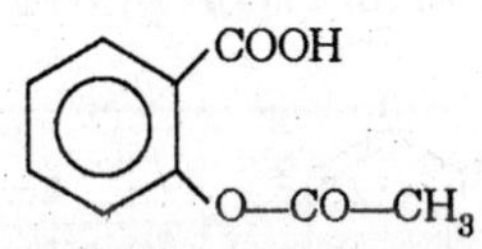

Aspirin (Acetyl salicylic Acid)

[vii] 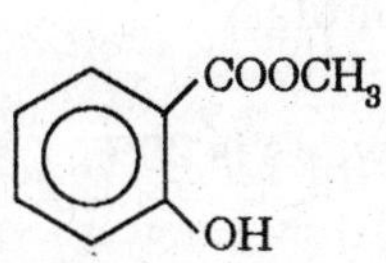

Oil of Wintergreen (Methyl Salicylate)

[viii] 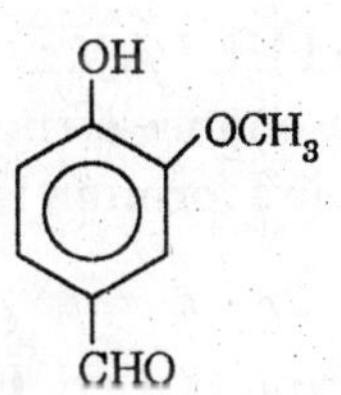

Vanillin

[ix] 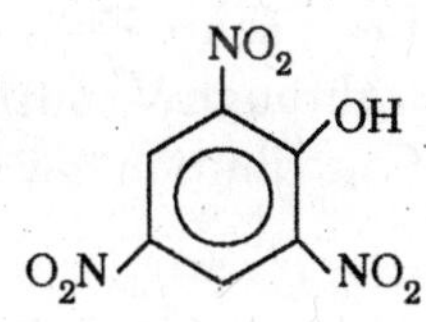

Picric Acid

[x] 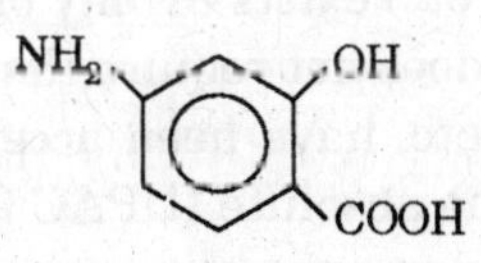

Para-Aminosalicylic Acid (PAS)

[xi] 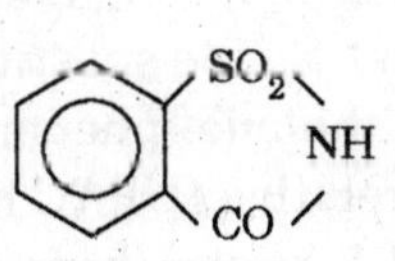

Saccharin

[xii] 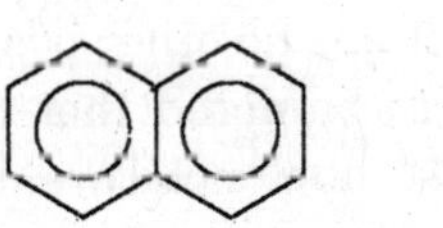

Naphthalene

[xiii] 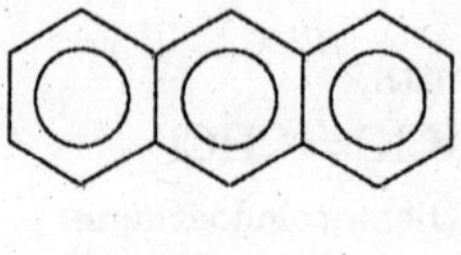 [xiv] 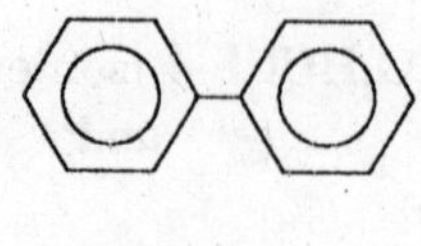 [xv] 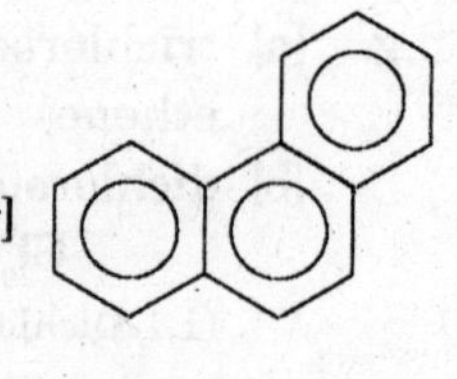

Anthracene Biphenyl Phenanthrene

Hundreds of aromatic compounds are known. They are considered as the derivatives of benzene. Benzene, C_6H_6, has the degree of unsaturation, D.U. = $\left(6+1-\frac{6}{2}\right)$ = 4, which indicates the minimum requirement of D.U. in aromatic compounds. It should be noted that the reverse is not true, that is any compound having D.U. = 4, may not be aromatic, thus, C_6H_6 = D.U. = 4 is  **or**  **and many other compounds are not aromatic.**

The single bond and double bond are continuously changing their positions inside the ring, so fast that they cannot be located at a particular position. This is called *resonance*, and each individual structure contributing to the resonance is called *canonical structure*. The combined effect (that is the resultant) of these canonical structures is called resonance hybrid and represented by the Robinson's sextet formula [A] above.

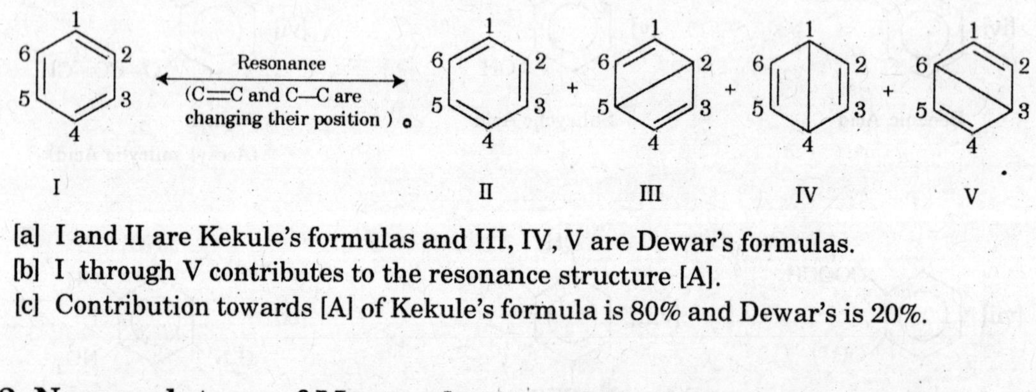

I II III IV V

[a] I and II are Kekule's formulas and III, IV, V are Dewar's formulas.

[b] I through V contributes to the resonance structure [A].

[c] Contribution towards [A] of Kekule's formula is 80% and Dewar's is 20%.

22. Nomenclature of Monosubstituted Benzene-compounds

Due to resonance the single bond and the double bond become identical in the resultant hybrid structure, hence all six H-atoms in benzene become identical. Therefore, monosubstituted benzene, C_6H_5Q (Q = halogens, OH, NO_2, alkyl group, etc) exists in only one form (i.e., it has no other positional isomers). So the substituent, Q, does not require the locant. It is to be noted that benzene, naphthalene, phenanthrene, etc. have been accepted in IUPAC nomenclature, but these are relative IUPAC names. The absolute IUPAC name of benzene appears to be cyclohexa–1,3,5–triene (hence easy to draw the structure of benzene).

Similarly,

[a]

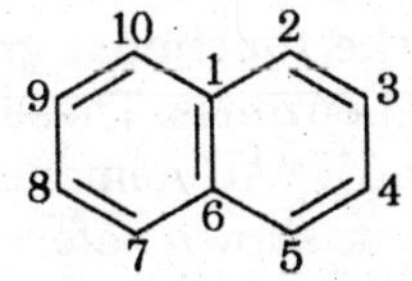

[i] Naphthalene (realtive name, common name) also IUPAC. (the locants are different, vide 24[a])

[ii] Bicyclo [4:4:0] deca 1(6), 2, 4, 7,9–pentaene (absolute IUPAC name and numbering).

[b]

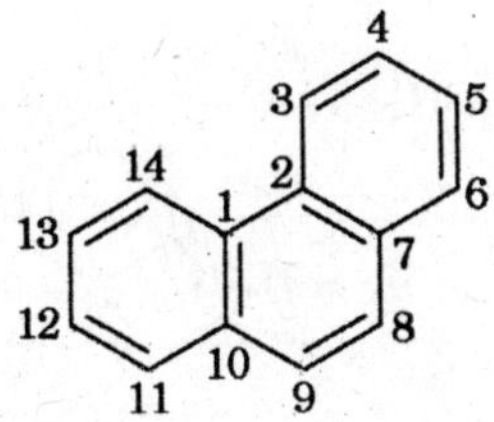

[i] Phenanthrene (relative name, common name, vide 24[d])

[ii] Tricyclo [8:4:0:0^{2,7}] tetradeca–1(10), 2(7), 3,5,8,11,13–heptaene (systematic name).

These two examples illustrate that the absolute IUPAC nomenclature of many compounds may be useful in drawing the structure of the molecule and predicting several chemical properties yet the common names of hundreds of compounds are used due to their simplicity (hence accepted in systematic naming).

Examples

Benzene

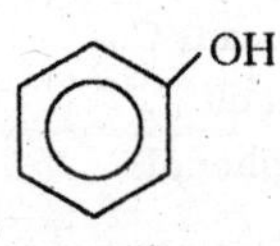

Hydroxybenzene (IUPAC)
Phenol (common name). Carbolic acid (commercial name)

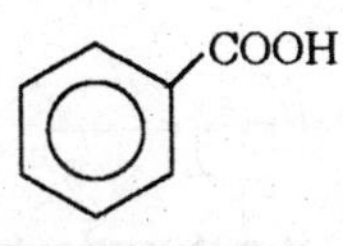

Carboxybenzene (IUPAC)
Benzoic Acid (common name) (commercial name)

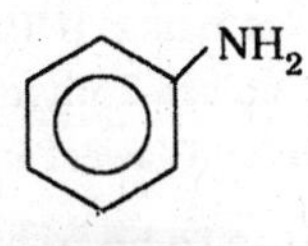

Aminobenzene (IUPAC)
Aniline (common name and commercial name)

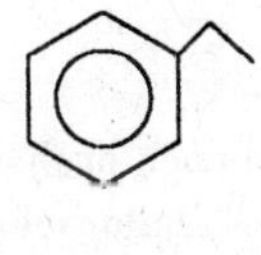

Ethylbenze

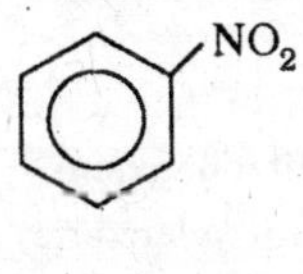

Nitrobenzene (IUPAC name and common name). Oil of mirbane (commercial name)

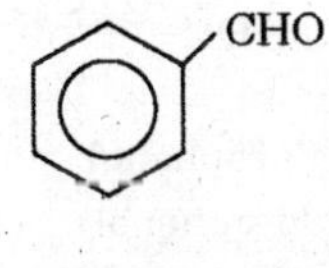

Benzaldehyde C_7H_6O
Oil of bitter almond (commercial name)

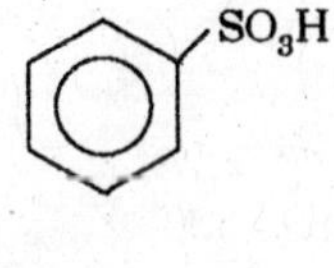

Benzene Sulphonic Acid, $C_6H_6O_3S$

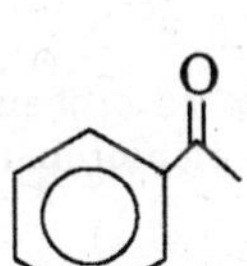

Acetophenone, C_8H_8O (common name), 1–oxo–1-phenylethane.

23. *Ortho*, *meta* and *para* (i.e., *o*, *m* and *p*) *notations*

Three positional isomers (carbon skeleton remains same but the functional groups, i.e., substituents have different locants) are possible in disubstituted benzenes. These are *ortho*, *meta* and *para* (abbreviated as *o*, *m* and *p*, always italicized). In IUPAC nomenclature the locants are specified (instead of *o*, *m* and *p*) by Arabic numerals as shown before.

Examples

[a] 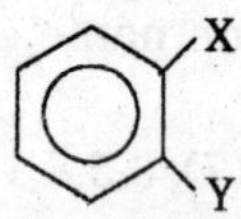 $X = Y = CH_3$ → 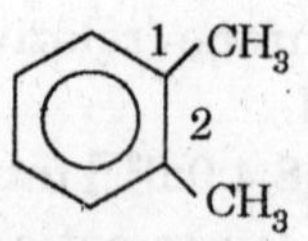 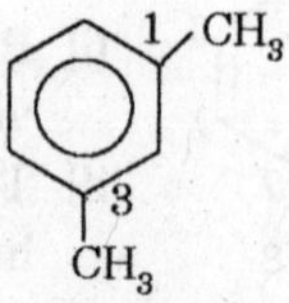 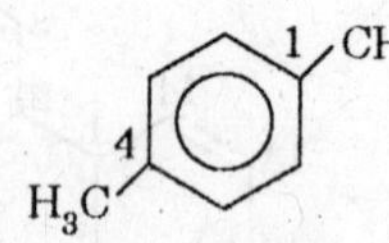

[X = Y or X ≠ Y]
Ortho-isomer (*o*)

o-Xylene
(IUPAC name:
1,2-dimethyl benzene)

m-Xylene
(1,3-dimethyl-
benzene)

p-Xylene
(1,4-dimethyl-
benzene)

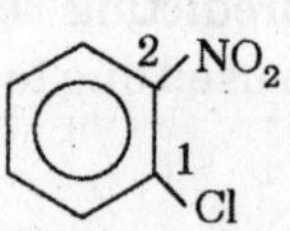

o-Chloronitrobenzene
(Alphabetic order; benzene root name; IUPAC name: 1-chloro-2-nitrobenzene; rule of lowest number)

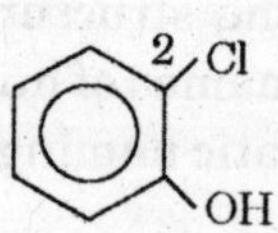

o-Chlorophenol
(phenol root name, OH has locant C_1. IUPAC name: 2-Chlorophenol)

It is to be noted that (*o*) if not underlined may be confused with an oxygen atom. So o stands for *ortho*-isomer. The circle inside the hexagon is not oxygen, but it is the group of six π-electrons (Robinson's Sextet).

[b] 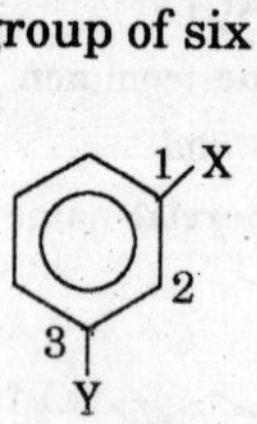 X = Cl, Y = OH → 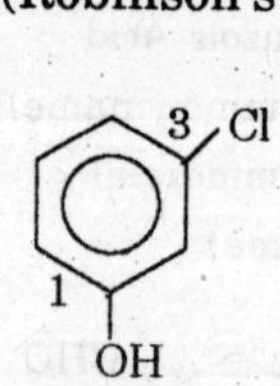 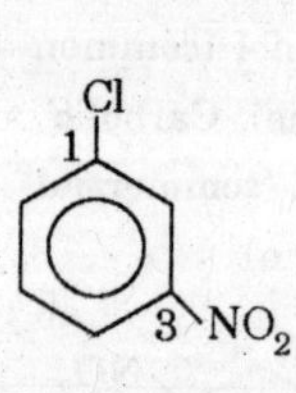 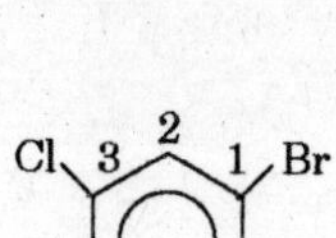

meta (*m*)
(1,3-locants)

m-chlorophenol
(3-Chlorophenol)
The suffix is at C_1.

m-chloronitrobenzene
(1-Chloro-3-nitrobenzene)

m-Bromochlorobenze
(1-Bromo-3-chlorobenzene)

[c] 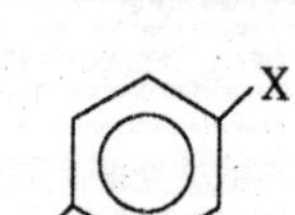 X = Cl, Y = OH → 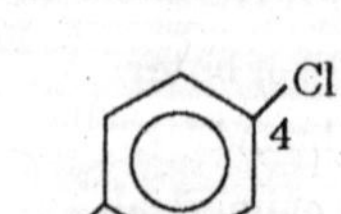 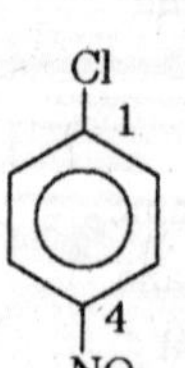 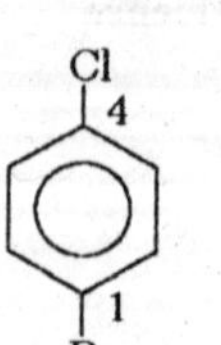

para (*p*)

p-Chlorophenol
(4-Chlorophenol)
The suffix is at C_1

p-Chloronitrobenze
(1-Chloro-4-nitrobenzene)
(Alphabetic order, starts with lowest number)

p-Bromochlorobenze
(1-Bromo-4-chlorobenzene)

[d] [i] Mono-substituted benzenes never furnish *o*, *m* and *p* (the question does not arise) isomers, hence mono-substituted benzenes have no isomers.

[ii] Any two adjacent positions in benzene is *ortho*. When only one carbon atom in the ring separates the two substituents, then the compound will be *meta* and two ring-carbons separating the two substituents will deliver the *para* compounds. Thus,

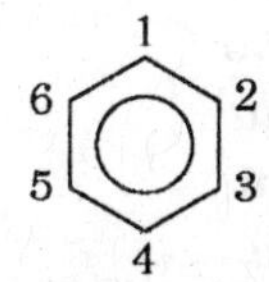

ortho: 1, 2 | 2, 3 | 3,4 | 4,5 | 5,6 | 6,1 are o to each other positions.

meta: 1,3 | 1,5 | 2,4 | 2,6 | 3,5 are *meta* to each other position.

para: 1,4 | 2,5 | 3,6 are *para* to each other positions.

[iii] Three substituted benzene compounds do not furnish the *ortho*, *meta* and *para* isomers (no question arises).

24. Nomenclature of Fused Ring Polycyclic Aromatics

When more than one aromatic rings are condensed with each other in the manner shown below (i.e., each ring connected with the other at adjacent carbons of each ring), then the fused ring aromatic compounds are generated. The following examples are illustrative.

[a]

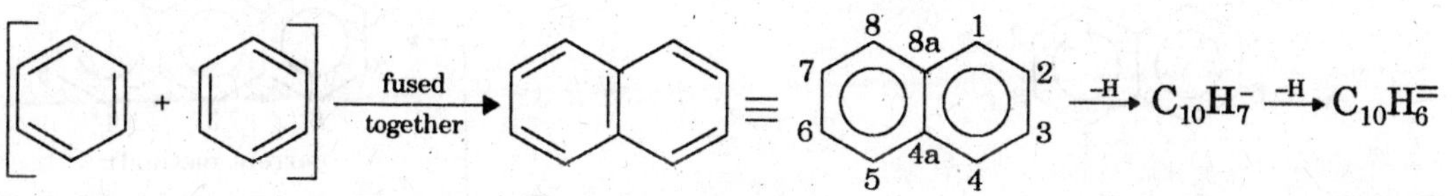

Benzene

Naphthalene ($C_{10}H_8$)

Naphthyl group (α and β)

Naphtho group (1,2 and 2,3)

The fusion points (or the ring junctions) are marked by English alphabet in lower case (i.e., 4a and 8a in naphthalene). The nearest carbon of the ring junctions is marked as C_1 and proceeded clockwise or anticlockwise along the skeleton of the compound. Also α and β notations are used to denote the nearest carbon and the next to the nearest carbon on both side of the ring juncture, respectively.

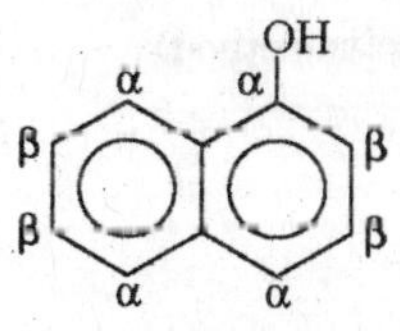

α-Naphthol (1–Naphthol, IUPAC)

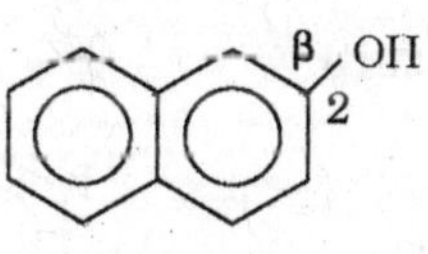

β-Naphthol (2–Naphthol, IUPAC)

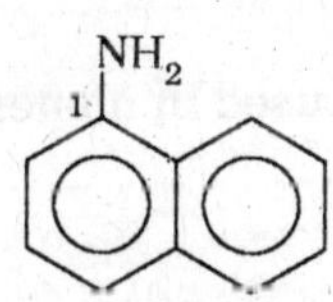

α-Naphthyl amine (1–Naphthylamine, IUPAC)

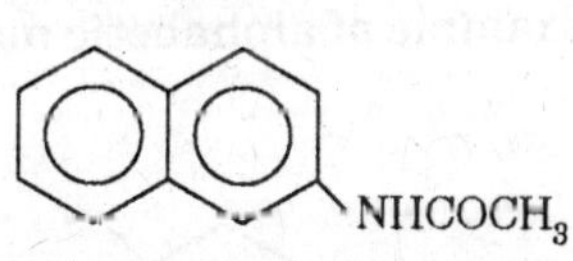

β-Naphthyl aminoacetate (2-Acetyl aminonaphthalene IUPAC)

It is to be noted that the following [A], [B] and similar compounds connected to each other at only one carbon atom of each ring are not fused ring compounds.

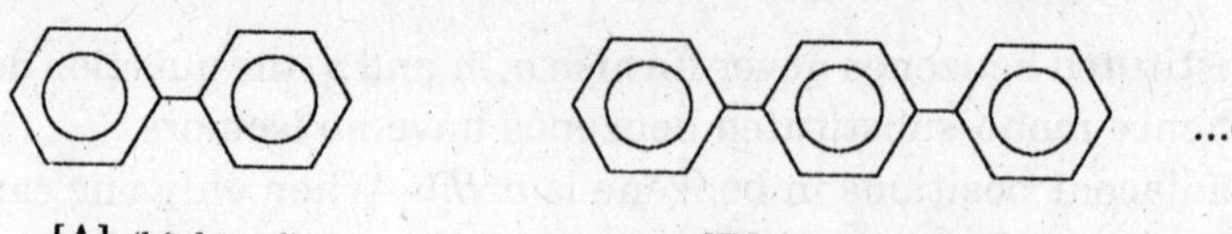

[A] (biphenyl) [B] (terphenyl)

[b] The relationships of different locants of naphthalene are shown below.

locants	names
1,2	*ortho*
1,3	*meta*
1,4	*para*
1,5	*ana*
1,6	*epi*
1,7	*kata*
1,8	*peri*
2,6	*amphi*
2,7	*pros*

[c] **Nomenclature of 'acene'-series**: In this series more than two aromatic rings are fused together side by side in a straight line. The first member is considered to be derived from naphthalene employing its $C_2—C_3$ bond for the fusion of the third ring as shown below.

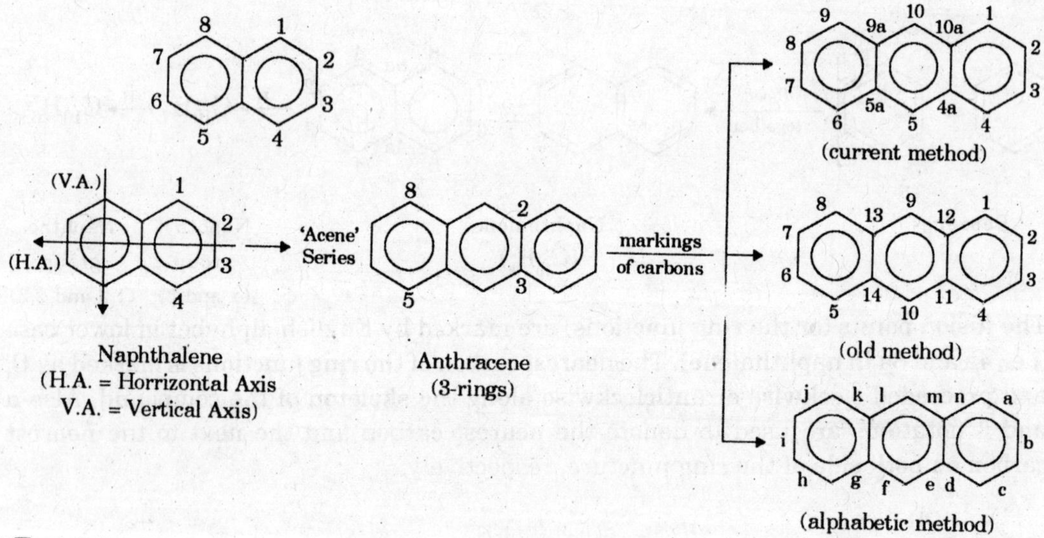

Example of alphabetic marking used in nomenclature:

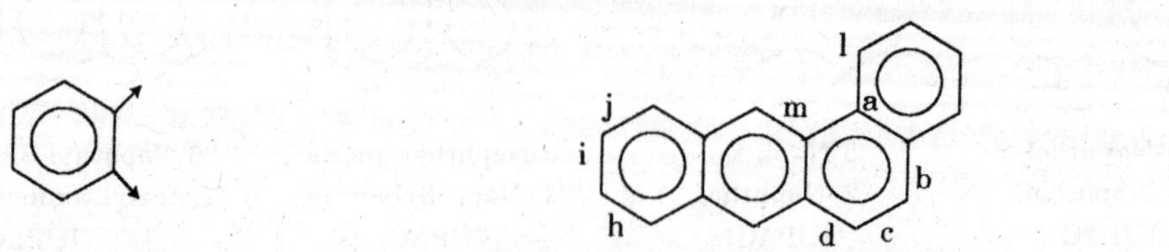

Benzo group, C_6H_4 = Benz [a] anthracene = Benzo [a] anthracene

(Benzo is attached to bond marked 'a' in anthracene)

It should be known that many polycyclic aromatic compounds can generate cancer (i.e. carcinogenic compounds). The nomenclatures of aromatic compounds are comparatively easy but the carbon frame and numbering of the frame should be known.

Examples

[i] 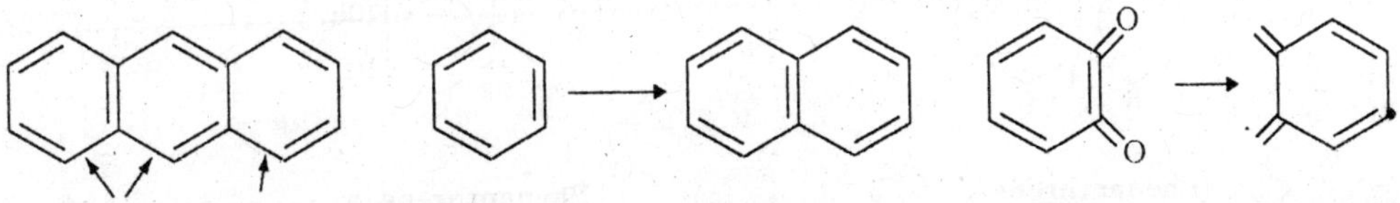

Benzenoid rings — Quinonoid ring — Benzene — Benzenoid (like Benzene) — Ortho quinone — Quinonoid (like quinone)

[ii] 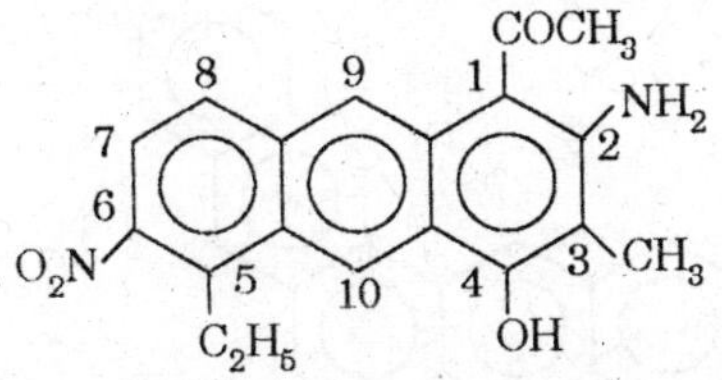

Old system of marking of carbon skeleton has been followed.

Nomenclature

[A] **The *root name* of the compound is anthracene.**

[B] **The carbons are marked (here old system has been followed).**

[C] **The substituents are arranged in the alphabetic order. Thus, Acetyl (Ac)–Amine (Am)–Ethyl (E)–Hydroxy (H)–Methyl (M)–Nitro (N).**

[D] **These substituents are used as prefixes with the corresponding locants using anthracene as the suffix.**

[E] **So the apparently complex molecule has the name: 1–Acetyl–2–amino–5–ethyl–4–hydroxy–3–methyl–6–nitroanthracene ('o' and 'a' both stay in 'oa').**

[F] **There are some exceptions to the [C] above, one example is methyl hydrogen phthalate (I) when 'm' comes before 'h'. The molecule is *not* named as hydrogen methyl phthalate.**

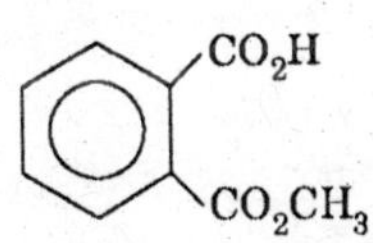

[I]

[iii] 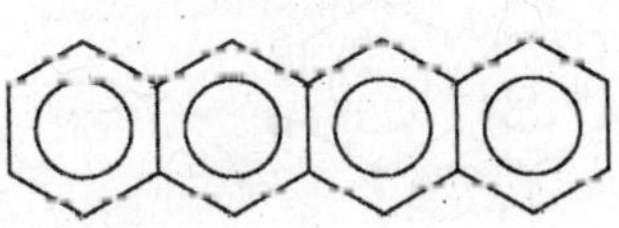

Tetracene
(Four rings fused in 'acene' series)

[iv] 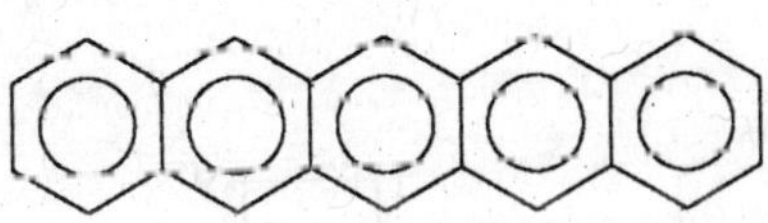 ... etc.

Pentacene
(Five rings in 'acene' series)

[d] **Nomenclature of 'phene'-series**: **The first member is phenanthrene, and in this series the parent skeleton of phenanthrene is present anywhere in the molecule.**

Examples

[i] 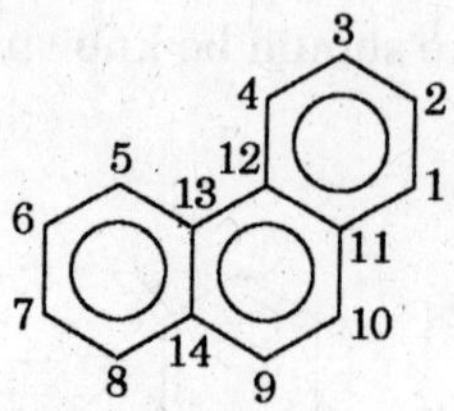

Phenanthrene
(old numbering)

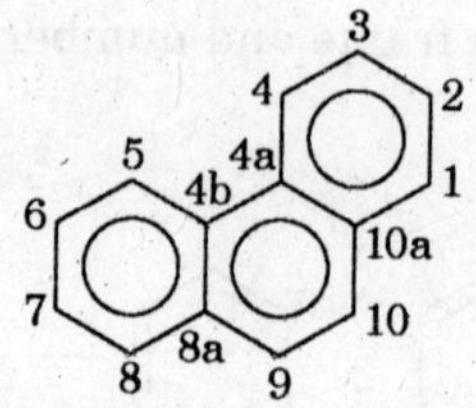

Phenanthrene
(curernt numbering)

[ii] 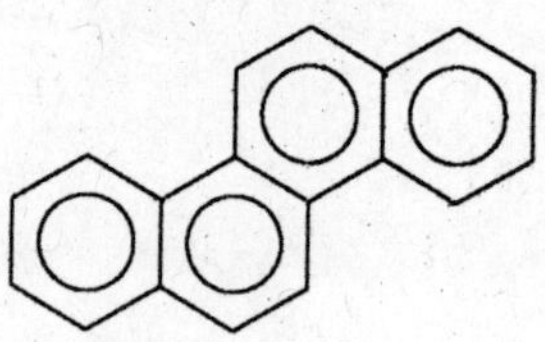

Chrysene
(Tetraphene)

[iii] 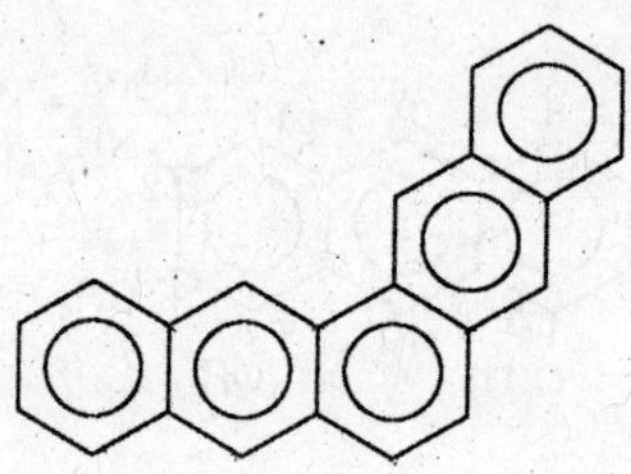

Pentaphene

[iv] 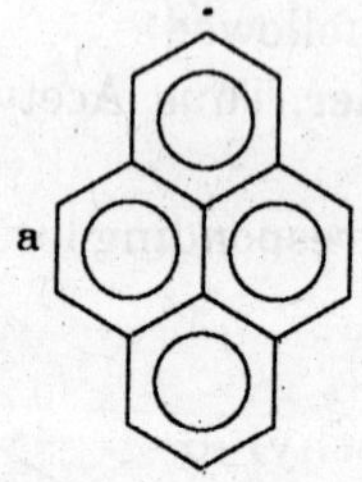

Pyrene

[v] 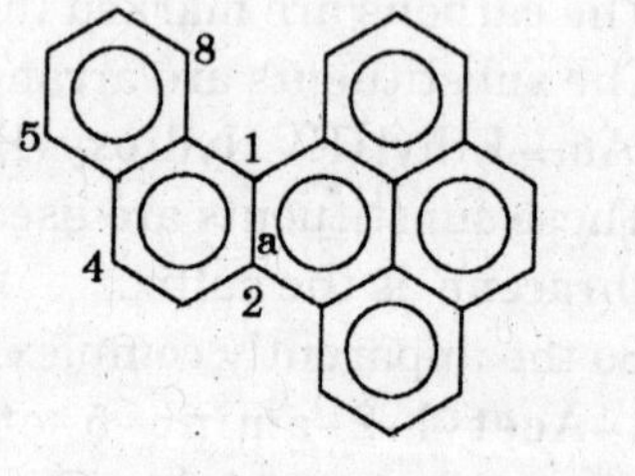

Naphtho [1, 2–a] pyrene

[vi] 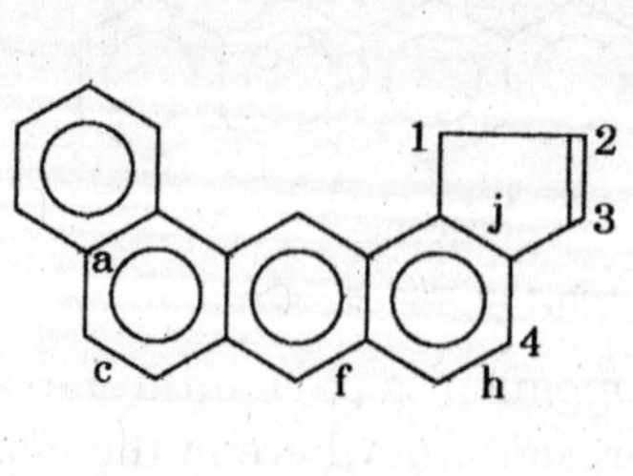

1*H*-Benzo [a] cyclopent [j] anthracene

[vii] 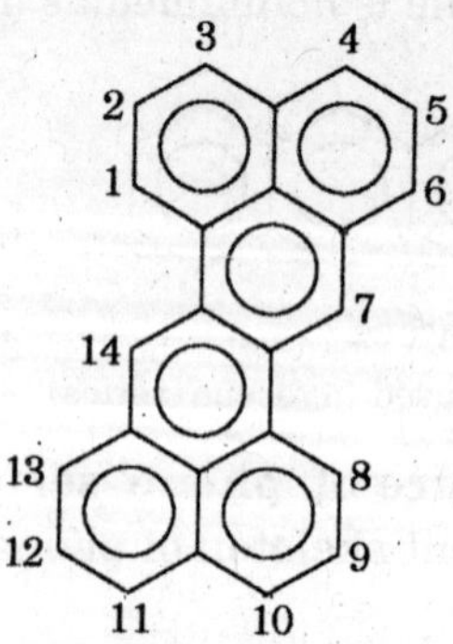

Zethrene

[viii]

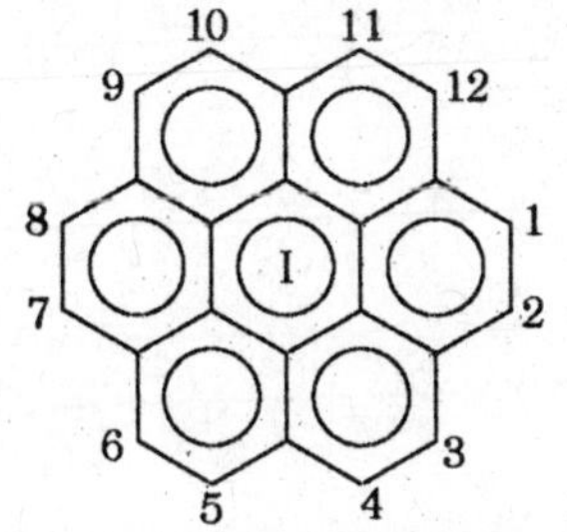

Circobenzene (also called coronene)

The central benzene ring (I) is circled by benzene rings .

[e] **Numbering of polycyclic aromatic hydrocarbons**: This is done by the following general procedure. With the horizontal and vertical axes four quadrants have been obtained as shown below.

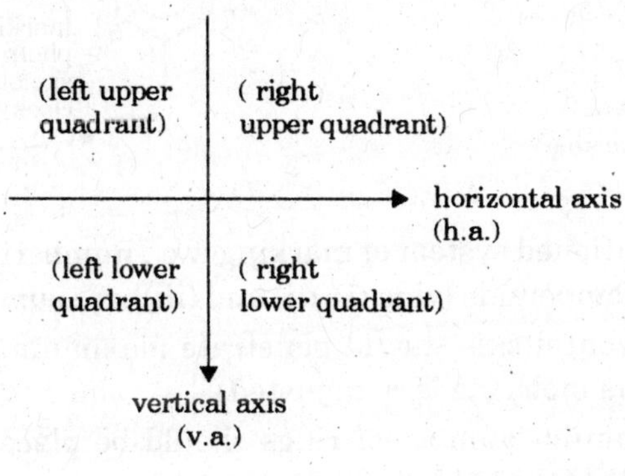

[i]

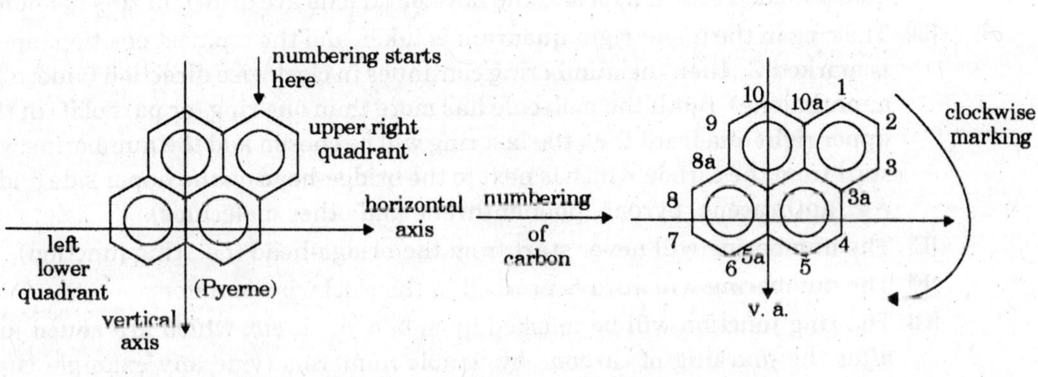

[ii] 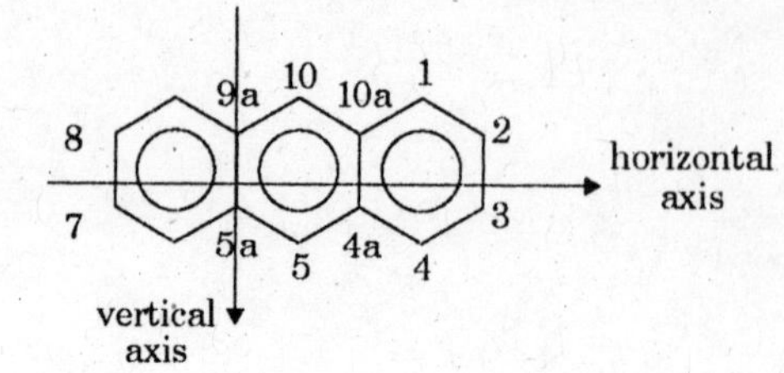

(marking of anthracene skeleton)

[iii] 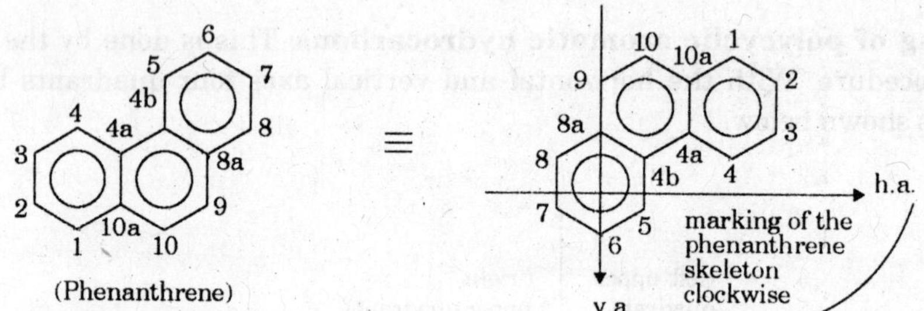

[iv] The aforementioned system of marking (i.e., numbering) the carbons of the skeleton of such compounds [vide (i), (ii) and (iii)] are summerised below.

[A] The horizontal axis should penetrate maximum number of rings (the skeleton of the molecule is so adjusted).

[B] The maximum number of rings should be placed in the upper right-hand quarter (the vertical axis is so adjusted).

[C] The minimum number of rings (or part of it) should be in the lower left-hand quarter (the vertical axis and the horizontal axis are drawn in this fashion).

[D] The ring in the upper right quadrant is taken and the topmost position (apex) is marked C_1, then the numbering continues in clockwise direction (vide, e.g., naphthalene). But if the molecule has more than one ring (or part of it) in the upper right quadrant then the last ring will be chosen and the numbering will start from the carbon which is next to the bridge-head on the upper side (vide-e.g., anthracene, pyrene, phenanthrene and other molecules).

[E] The numbering will never start from the bridge-head (i.e., ring junction).

[F] The numbering will always proceed in the clockwise manner.

[G] The ring junction will be marked by a, b, c, d, ..., etc, which are added just after the marking of carbons by Arabic numerals (vide any example cited before, particularly benzanthracene and naphthophenanthrene).

25. Examples of a Few Interesting Cyclic Compounds

[A] [i] 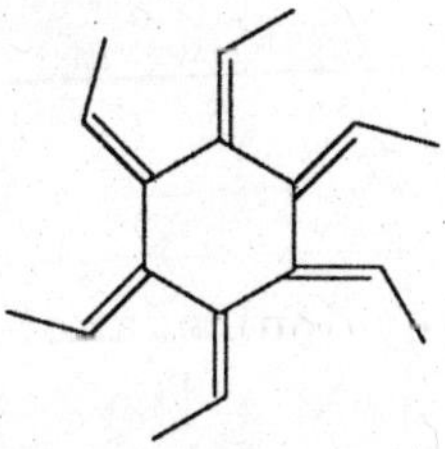

[ii] 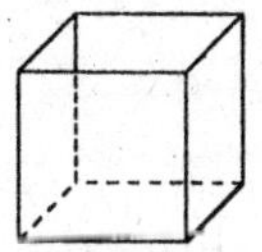

[iii] 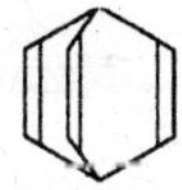

Radialene $C_{18}H_{24}$

Cubane C_8H_8

Barrelene C_8H_8

[iv] 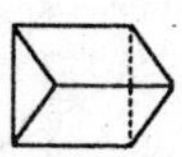

[v] 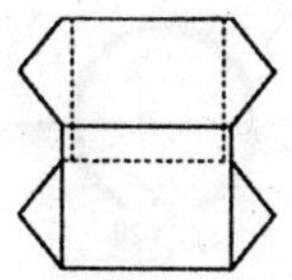

Prismane (C_6H_6)

Asterane ($C_{12}H_{16}$)

[B] **Catenanes, Rotaxanes and Knots**: A few examples of these interesting and less well known compounds are given below.

[a] Catenanes

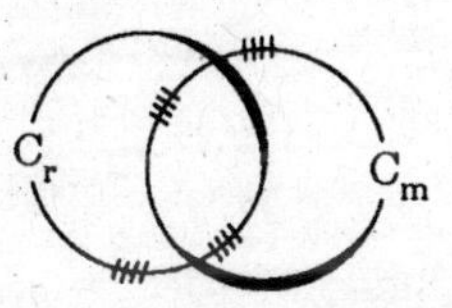

[i] 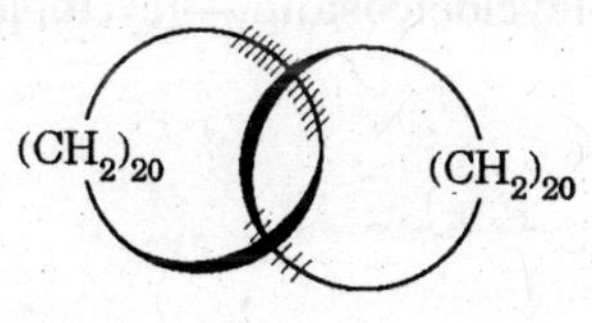

Name: [2]—[Cycloeicosane]—[cycloeicosane]—catcnane

- dash (↓): the dash after [2]
- dash (↓): the dash after [Cycloeicosane]
- dash (↓): the dash before catcnane
- [2] (↑): No. of rings in the Catenano
- [Cycloeicosane] (↑): No. of carbons in the first ring. Cyclo prefix.
- [cycloeicosane] (↑): No. of carbons in the second ring. Cyclo prefix.
- catcnane (↑): Class of compound

[ii] 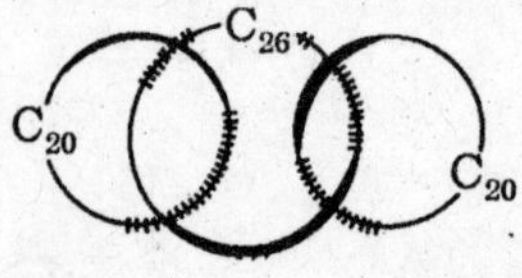

Name:

[3]—[Cycloeicosane]—[cyclohexacosane]—[cycloeicosane]—catenane

↑ No. of rings ↑ 1st ring. ↑ middle ring (no. of carbon atoms) ↑ 3rd ring

[b] Rotaxane: The examples are shown below.

[i] 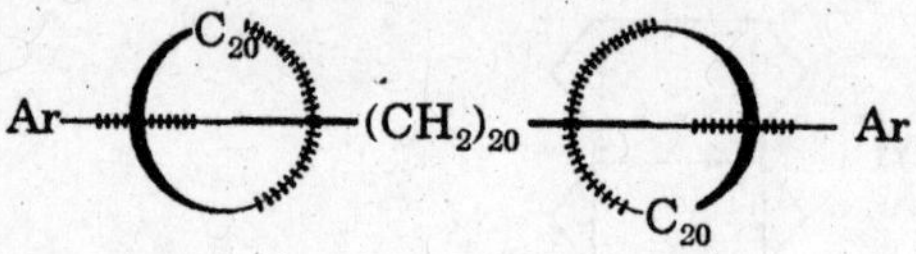

Name:

[3]—[1,20-Diaryleicosane]—[cycloeicosane]—[cycloeicosane]—rotaxane

↑ Three units are present ↑↑ locants of two aryl groups ↑ chain length ↑ dash ↑ 1st ring ↑ 2nd ring ↑ dash ↑ class

[ii] 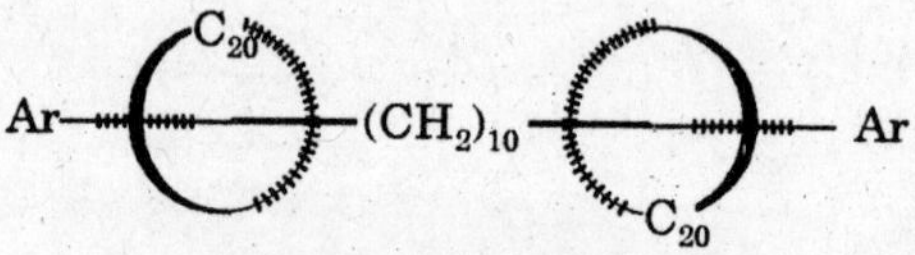

Name: [3]—[1,10-Diaryldecane]—[cycloeicosane]—[cycloeicosane]—rotaxane

[iii] 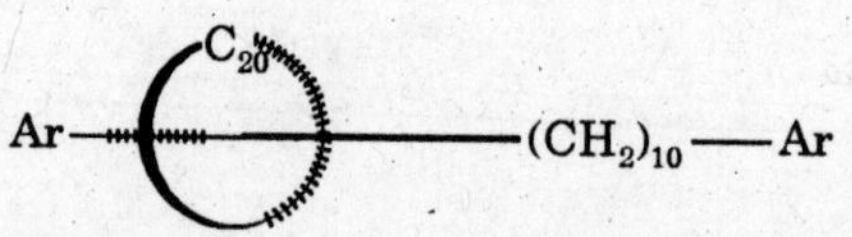

[2]–[1,10-Diaryldecane]–[cycloeicosane]–rotaxane

[c] Knots: One example is given below:

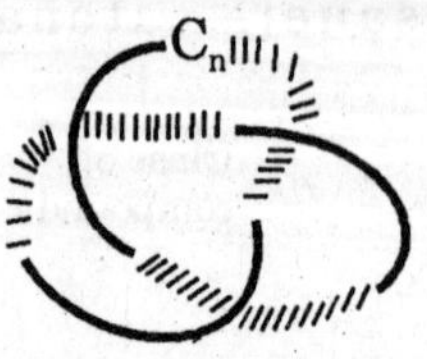

26. Annulenes

When in a compound having only one cycle there are alternate single bond and double bond (C—C and C=C), then the compound will be termed annulene. When the number of double bonds in the cycle is '*n*' (*n*=2, 3, 4, ..., etc.), then the total number of carbon atoms in the cycle will be '2*n*' which will be denoted as [2*n*]. Thus [4] annulene will be cyclobutadiene [I], [6] annulene is benzene (aromatic), [8] annulene is cyclooctatetraene (no vowel is cancelled), this is *not planar*, (hence not aromatic) (II), [10] annulene (*not* aromatic, because non planar) [III], but 1,6-methano [10] annulene [IV] is aromatic because it is planar. Hence it is clear that [i] cycle, [ii] conjugated single bond and double bond, [iii] planarity, and [iv] high resonance energy are the chief criteria of a compound to be recognised as an aromatic compound.

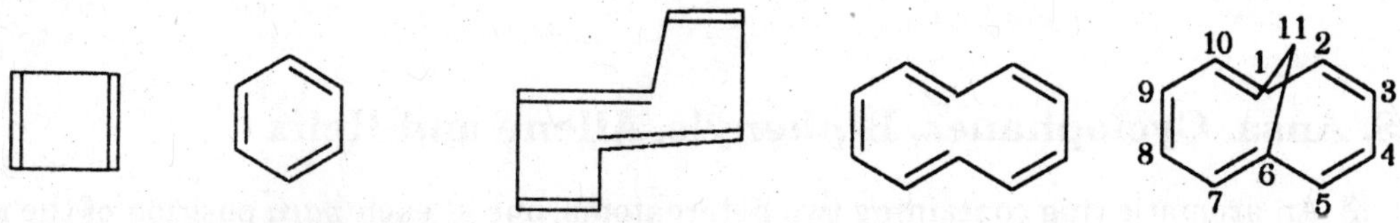

[I] (Benzene) [II] [III] [IV]

[14] annulene and [18] annulene are aromatic following Hückel's rule, (4*n*+2)π-electrons are present in planar molecule, but [12] annulene and [16] annulene are not aromatic compounds.

27. Metallocene

It has been observed that a particular compund having aromatic property has special aptitude to combine with a special metal ion to make stable compound where the metal ion is arrested as if in a cage. Such complex compounds are called metallocene. Thus,

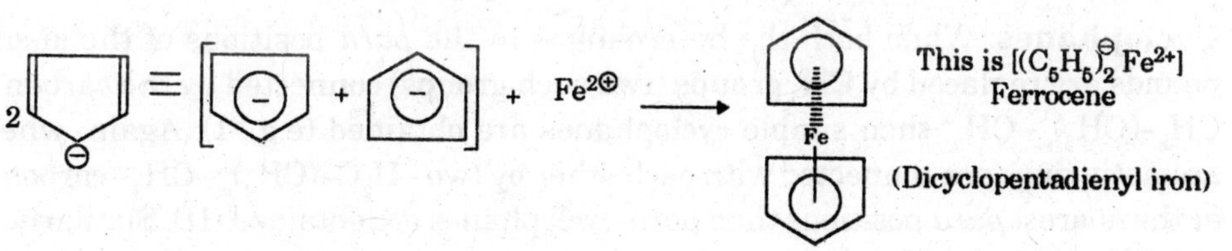

28. Charge-transfer Compounds and π-complexes

When an electron donor organic compound combines with an electron acceptor to form a complex compound then a charge-transfer compound is formed. Thus,

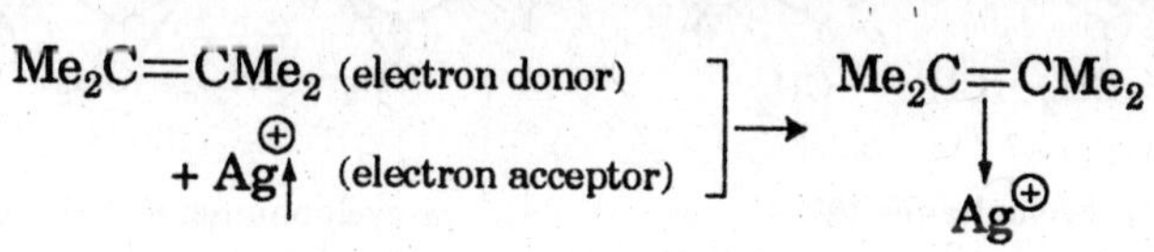

Again when the donor molecule is rich in π-electrons and it combines with an electron acceptor, then generally the π-complex is formed. The donor molecule does not lose its aromatic character during the production of π-complexes. Thus,

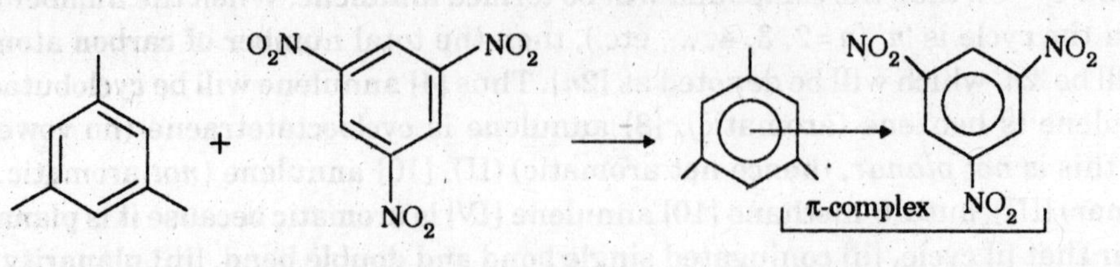

Electron donor (an aromatic hydrocarbon) **Electron acceptor** (also aromatic) **Trinitrobenzene derivative** (TNB-derivative)

29. Ansa, Cyclophanes, Biphenyls, Allene and Helix

[i] An aromatic ring containing two heteroatoms, one at each *para* position of the ring, when connected by $(CH_2)_m$-chain will constitute the ansa compound. Ansa means handle. Sometimes properly substituted ansa compounds may exhibit optical activity. Thus,

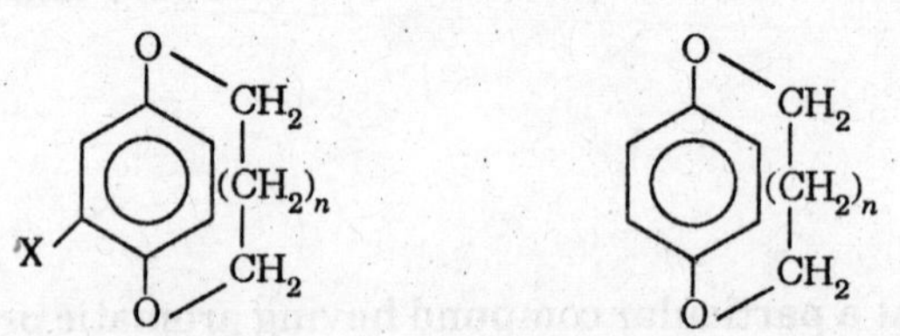

Ansa (X≠H); (optically active) Ansa (opacity inactive)

[ii] **Cyclophanes**: When both the heteroatoms in the *para* positions of the ansa compounds are replaced by CH_2 groups (two such groups) connected by the carbon chain CH_2–$(CH_2)_m$–CH_2, then simple cyclophanes are obtained (e.g., I). Again, when two aromatic rings are connected with each other by two –H_2C–$(CH_2)_m$–CH_2– carbon chain at the nearest *para* positions then *para*-cyclophanes are obtained (II). Similarly, when two CH_2–$(CH_2)_m$–CH_2– chains connect the *meta* positions, then meta-cyclophanes (III) are obtained.

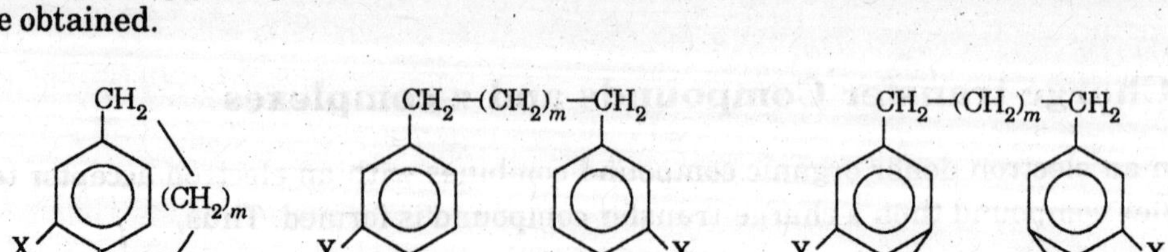

Simple cyclophane (I) *p*-cyclophanes (II) *m*-cyclophanes (III)

[iii] **Biphenyls**: These are formed when two phenyl groups (C_6H_5) are connected to each other by a single bond. They are different from fused ring aromatic compounds like naphthalene. The multiplying prefix (viz. bi, ter, tetra, etc.) varies with the number of phenyl rings connected by single bonds. The numbering of carbons are shown below.

[a] 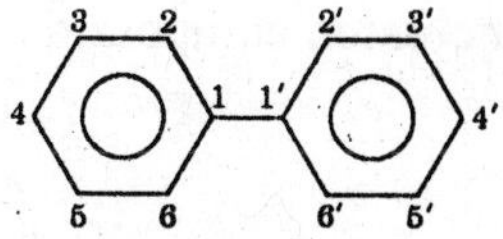

Simple biphenyl

[b] 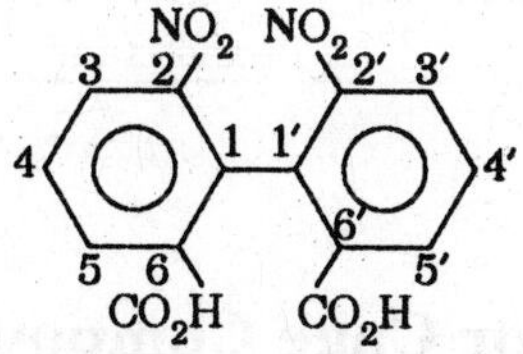

Optically active substituted biphenyl
(*d* and *l*)

[iv] **Allenes**: These compounds have the skeleton C=C=C [i.e., *n*, (*n*+1)-diene system] and they may be optically active with proper substituents; abC=C_n=Cde (*n* >1 and even /odd) is cumulene.

[a] 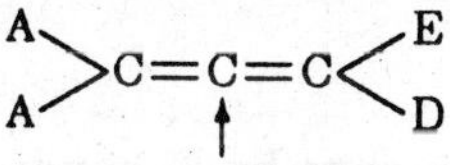

Optically inactive
allene ($n=1$)

[b] 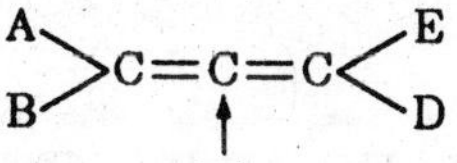

Optically active allene
($n=1$)

[c] 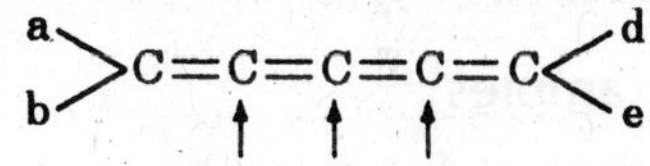

(Cumulene, $n=3$)

[v] **Helix**: The example is shown below: the carbon skeleton is the form of a helix.

Helix

30. Inclusion Compounds

When in the crystal lattice of an organic compound a tunnel or a cave like space is available, then a different type of compound can enter there and stay. Such compounds are named 'inclusion' molecules where the guest molecule is stabilised by the van der Waals forces of the host molecules. If the size of the guest molecule becomes larger than the tunnel or space of the host, then the guest cannot enter. Again, if the tunnel is larger than the size of the guest, then the latter comes out of cavity very easily. So the size of the guest molecule and the size of the space or tunnel of the host molecule must match each other to make the moderately stable inclusion compound.

Examples

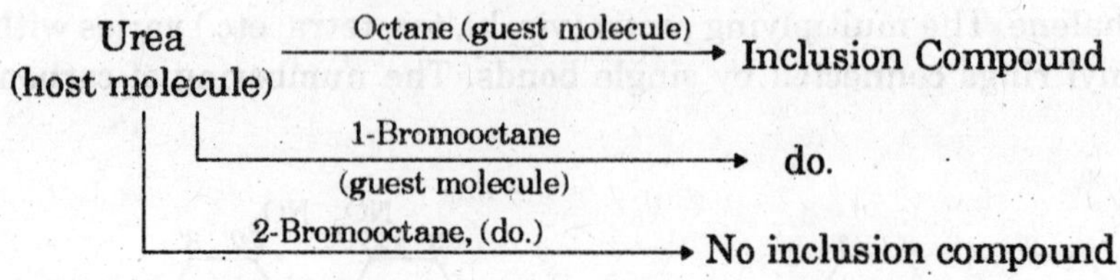

31. Clathrates or Cage Compounds

If an organic compound has a cage like structure of the crystal, then in the vacant space of the crystal a molecule of other compound (the size must be proper) can enter and be arrested there like a bird in a cage. Such complex compound derived from the cage-crystal and the guest molecule is known as clathrates or cage compounds which are stable at low temprature but at higher temperature as soon as the crystal melts the clathrate decomposes.

Examples

Three molecules of hydroquinone (*p*-HO—C_6H_4—OH) make a cage by H-bonding. In the vacant space inside the cage a molecule of methanol can enter and get arrested to make the cage compound, i.e., clathrate. When methanol is replaced by ethanol, then the clathrate is not formed because the larger ethanol (compared to methanol) cannot be accommodated inside the cage.

32. Crown Ether Complexes and Cryptates

The general formula of crown ether is [*x*]-Crown-*y*, ($x > y$). Total number of members of the cycle is x and the number of oxygen atoms (included in x) is denoted by y. So $(x - y)$ = number of carbons in the crown ether molecule. Generally, $x = 3y$ is observed. The cavity inside the crown ether can arrest an atom of alkali metal ion (viz., Li^+, Na^+, K^+, etc) depending on the sizes of the cavity and the metal ions. This property has been utilised in dissolving many inorganic compounds in crown ether. The anhydrous organic solvents cannot dissolve these water soluble inorganic reagents. Hence during performing the reaction in absolutely anhydrous condition the crown ethers with proper cavity are utilised. This has proved itself to be a very interesting synthetic procedure in organic chemistry. But the crown ethers are expensive solvents hence their uses are limited.

Examples

[i] **[12]-crown-4[A], $(CH_2)_8O_4$ [here $x = 12$ and $y = 4$]**
The cavity inside the molecule is small. So only $Li^{\oplus}$ (small sized ion) can be arrested because $Li^{\oplus}$ can enter into the molecule and captured by lone pairs of electrons of oxygen. $Na^{\oplus}$ or $K^{\oplus}$ cannot enter into the cavity due to larger sizes compared with $Li^{\oplus}$.

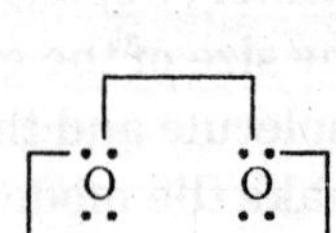

[A]

[ii] [15]-crown-5, [B]; [x=15, y=5; CH_2 = (15 – 5) =10]; $C_{10}H_{20}O_5$.
The structural formula of (B) is drawn in the following way. To start with the oxygen atoms (5 in this case) are drawn symmetrically. Each one of them is connected to the other through $-CH_2-CH_2-$ groups. In this crown ether $Na_2Cr_2O_7$ dissolves in anhydrous state but $KMnO_4$ does not dissolve in [15]-crown-5 because $Na^{\oplus}$ can enter the cavity which is just of right dimension of arresting the $Na^{\oplus}$ ion while $K^{\oplus}$ is too big for the cavity which does not permit the entry of larger $K^{\oplus}$

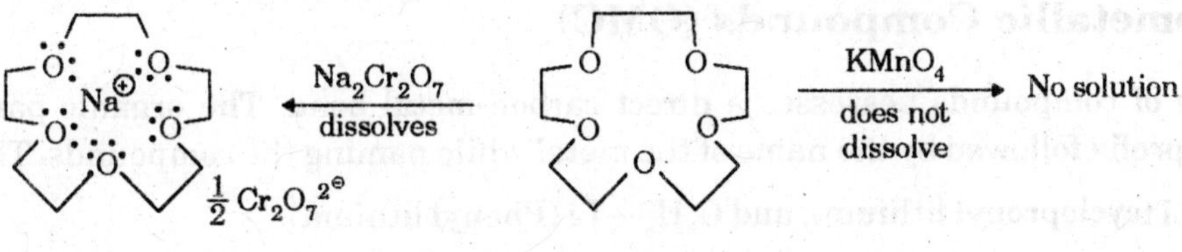

[B]

[iii] [18]-crown-6, [C]; [x=18, y= 6; (x–y)=12=$(CH_2)_{12}$]; $C_{12}H_{24}O_6$.
The cavity inside the ether [C] can allow the entry of K^+ (ionic diameter: 2.66 Å) and hold it there. But Li^+ (ionic diameter: 1.20 Å) can enter but easily comes out of the cavity because of the smaller ionic diameter.

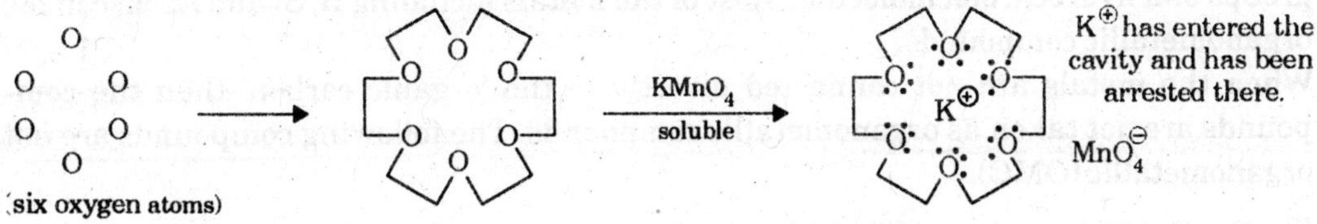

(six oxygen atoms)

[C]

[iv] **Dibenzo-[18]-crown-6 [D] and Dicyclohexyl-[18]-crown-6 [E].**

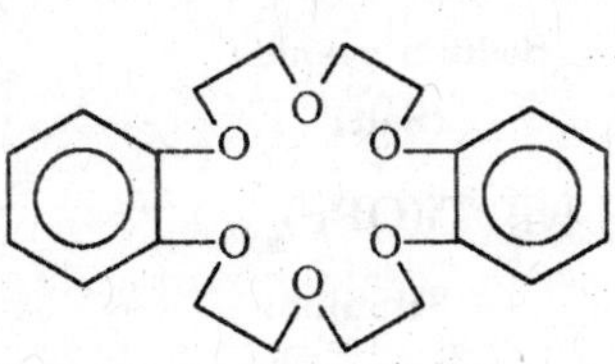

[D]

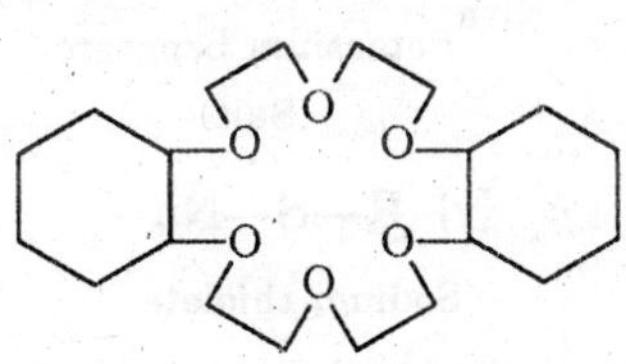

[E]

[v] **Cryptands: When the bicyclo crown ethers contain both oxygen and heteroatoms (N, S, etc.), then cryptands are formed which can arrest the metal ions inside to generate the cryptates.**

Examples

[i]

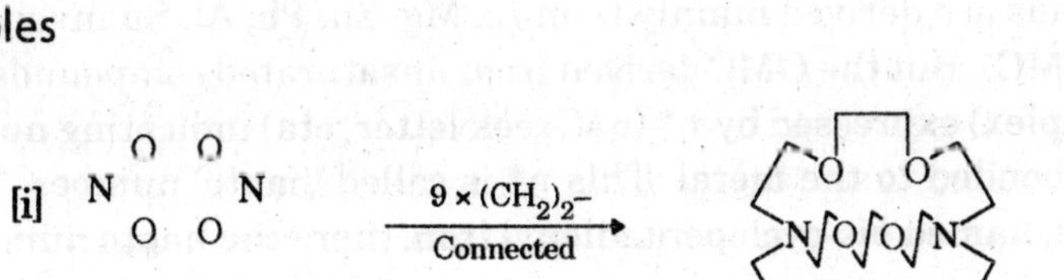

Cryptand; $[(CH_2)_9N_2O_4S_2]$

[ii] 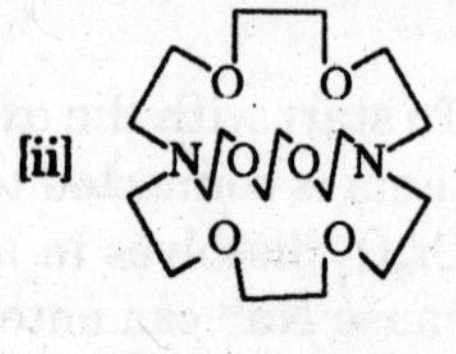

Kryptofix-222; $[(CH_2)_9N_2O_6]$

33. Organometallic Compounds (OMC)

[a] This class of compounds possesses a direct carbon-metal bond. The organic part is made the prefix followed by the name of the metal while naming the compounds. Thus,

[i] ▷—Li (cyclopropyl lithium), and C_6H_5—Li (Phenyl lithium)

[ii] CH_2=CH—Na (vinyl sodium); CH≡C—Na (ethynyl sodium or sodium acetylide); like a binary compound.

[iii] CH_3—Mg—I (Methyl magnesium iodide; Grignard reagent).

[iv] Me_2AlCl (Dimethyl aluminium chloride)

[v] HC≡C—MgBr (Ethynyl magnesium bromide), etc.

All these compounds have C*—metal bonds. Here C* stands for alkyl, aryl, cycloalkyl groups or a hydrocarbon molecule. Most of the metals including B, Si and As also make organometallic compounds.

[b] When the metals are not connected directly to the 'organic carbon' then the compounds are not taken as organometallic compounds. The following compounds are not organometallic (OMC).

[i] R—C(=O)—ONa	[ii] C_6H_5—COOK	[iii] Na—C≡N
Sodium alkanoate (Salt)	Potassium benzoate (Salt)	Sodium cyanide (Salt)
[iv] R—ONa	[v] R—S—Na	[vi] $Ti(OPr^i)_3$
Sodium alkoxide	Sodium thiolate	Titanium *iso*-propoxide

In this case the metal is named first, then the organic part follows. The name of N≡C—Na is noteworthy. In both (a) and (b), the metal part is always separated from the organic part (i.e., they do not make a single word). Thus,

[vii] Me—Zn—Me is dimethyl zinc, [viii] $(C_2H_5)_4Pb$ is tetraethyl lead.

[c] Organometallic compounds are derived mainly from Li, Mg, Zn, Pb, Al, Sn involving σ bond (called σ–bonded OMC). But the OMC derived from unsaturated compounds generates π-complex (pi-complex) expressed by η* (η=Greek letter, eta) indicating number of carbons (shown as x) bonded to the metal. This η* is called 'hapto' number. Thus, $(C_5H_5)_2Fe$ is $(\eta^5\text{–}C_5H_5)_2Fe$, named *bis*-cyclopentadienyl iron, (here the hapto number is 5)= Ferrocene.

Other examples of hapto numbers: M = An atom of the metal, η^x = (hapto ex)

[i] M—R

η^1 (*pr.* eta one)
(hapto number one)

[ii] R—Mg—I

η^1 (eta one)

[iii] 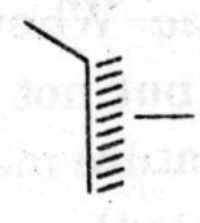

π-allyl
η^2 (eta two)

[iv] M— 

π-allene
η^3 (eta three)

[v] 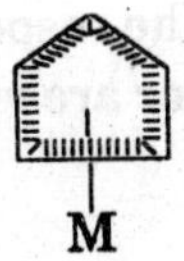

π-complex (η^5)
(η^x, x=5)

[vi] M— 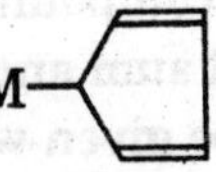

σ-complex (η')

[vii] M— 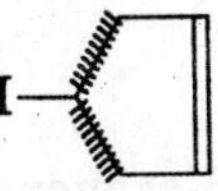 (hapto number three)

π-complex (η^3)

[viii] $K[PtCl_3(\eta^2\text{–}C_2H_4)]$ KCl = Zeise's salt, [ix] $Cr(\eta^6\text{–}C_6H_6)_2$ = Dibenzene chromium

[d] **Metal carbonyls**: These compounds retain both π- and σ-characters and general formula is $M(CO)_x$ [e.g., $Ni(CO)_4$=Nickel carbonyl; $Fe(CO)_5$ = Iron carbonyl; $Cr(CO)_6$ = Chromium carbonyl].

[e] **Ionic organometallic compounds**: M=metal and C=carbanion species, and $M^{\oplus} > C^{\ominus}$ (during naming the compounds).

Examples

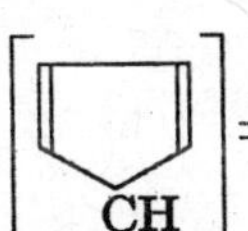 $= C_5H_5^{\ominus}$. Ionic OMC is $\overset{\oplus}{Na}\overset{\ominus}{C}\equiv\overset{\ominus}{C}Na^{\oplus}$ and $\overset{\oplus}{Na}(C_5H_5)^{\ominus}$.

[f] The organometallic compounds have been classified into three categories, viz., (i) symmetrical (and simple) when the alkyl groups (if more than one is present) are same (e.g., Et_2Zn, Et_4Pb, etc), (ii) unsymmetrical (and simple) when more than one alkyl groups are present and they are different (e.g., $EtHgC_3H_7$) and (iii) mixed, when the metal is connected to both organic part (i.e., alkyl group) and inorganic part (derived from inorganic acidic compounds), e.g., MeMg–I, Et_2SnCl_2 etc. In these examples the alkyl groups may be replaced by aryl groups.

34. Nomenclatures of Compounds containing Nitrogen, Oxygen and Sulphur

[i] Use of aza, oxa and thia prefixes to name the secondary and tertiary amines, ethers and thioethers, respectively, particularly when they are complex.

[a] ***Use of aza*: When N is present between two carbon atoms generating 2° and 3° amines (but not 1°-amines), then the aza prefix is used. This helps the IUPAC nomenclature making the understanding easier. The root name will be azaalkane (single word).**

At first the N is considered as carbon of the main chain to determine the root name. Then according to the rules applied in nomenclature of aliphatic and aromatic compounds the prefixes with the respective locants are selected. The rule of lowest sum and rule of lowest number are valid and the lowest number to the aza is to be given when possible.

Examples

[i] $\overset{3}{\ce{CH3-(CH2)2-NH}}-\overset{2}{\ce{CH}}(\ce{CH3})-\overset{1}{\ce{CH3}}$ = **2-Methyl-3-azahexane (*not* 5-methyl-4-azahexane). (not 3-Aza-2-methylhexane)**

[ii] $C_2H_5-CH(CH_3)-NH-C_2H_5$ = **3-Methyl-4-azahexane (*not* 4-methyl-3-azahexane cf '4s') (Rule of lowest sum)**

[iii] $Et-CH(CH_3)-N(Me)Et$ = **3,4-Dimethyl-3-azahexane (*not* 3,4-dimethyl-4-azahexane)**

[iv] 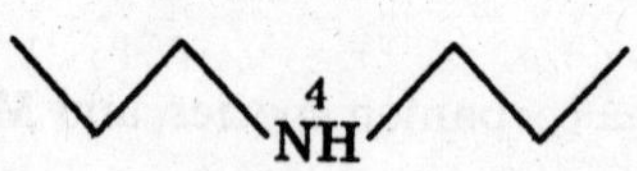 **= 4-Azaheptane**

[v] $\overset{1}{CH_3}-\overset{2}{NH}-\overset{3}{CH_2}-\overset{4}{NH}-\overset{5,6}{C_2H_5}$ = **2,4-Diazahexane (*not* 3,5-Diazahexane)**

[vi] 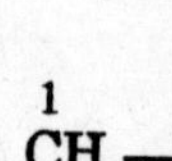 **= (Pyridine) = Azabenzene (locant absent)**

[vii] 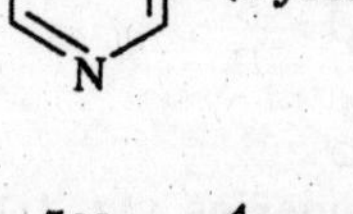 **= 1,3,5-Triazanaphthalene (cf.**  **naphthalene)**

[viii] 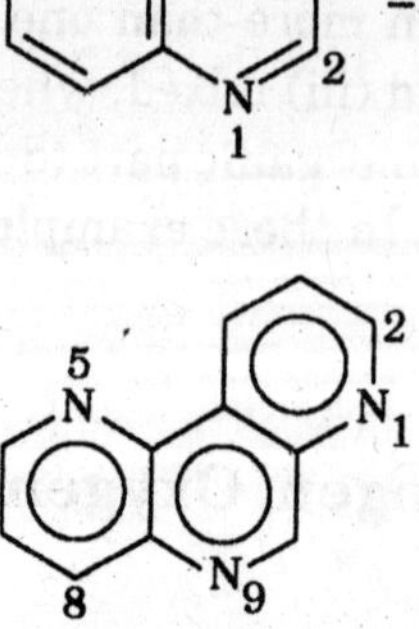 **= 1,5,9-Triazaphenanthrene (cf.** 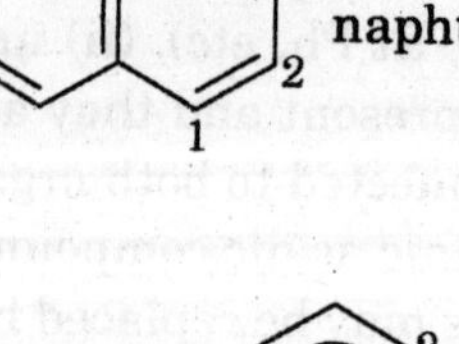 **phenanthrene)**

(*not* 4,8,10-triazaphenanthrene)

[ix] 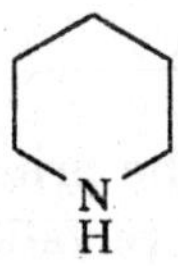 = (Piperidine) = Azacyclohexane (locant absent)

[x] 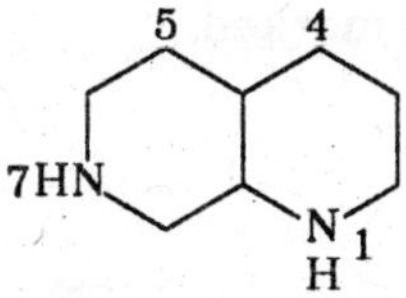 = 1,7-Diazadecalin (*not* 3,5-Diazadecalin, cf. '4s').

[b] When oxygen is present in the carbon chain as ether function (but not as other functional groups like alcohol, ester, carbonyl, etc.) and the molecule has complex structure, then oxa prefix (*not* oxo, as oxo stands for the carbonyl functional groups) is used. The rules for IUPAC naming of compounds are also valid in oxa nomenclature. The oxaalkane will be used as the single word.

Examples

[i] Et—CH_2—O—CH(Me)CH_3 = 2-Methyl-3-oxahexane (propyl *iso* propyl ether) (*not* 4-methyl-3-oxahexane)

[ii] CH_3—CH_2—CH(Me)—O—CH_2—CH_3 = 3-Methyl-4-oxahexane (cf. 4s) (*not* 4-methyl-3-oxahexane *or* 4-Oxa-3-methylhexane)

[iii] $\text{Me}\overset{2}{\text{O}}\text{—}(CH_2)_2\text{—}\overset{5}{\text{O}}\text{—}(CH_2)_2\text{—}\overset{8}{\text{O}}\text{—}CH_3$ = 2,5,8-Trioxanonane

[iv] 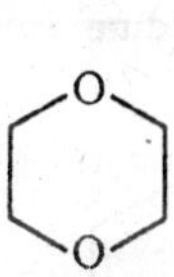 (Dioxane) = 1,4-Dioxacyclohexane.

[c] When sulphur replaces oxygen in ether, the compound is thioether and the prefix thia is used in such cases.

Examples

[i] $\text{Me}\overset{2}{\text{S}}\text{—}(CH_2)_2\text{—}\overset{5}{\text{S}}\text{—}CH_2\text{—}CH_3$ = 2,5-Dithiaheptane

[ii]  = 1,4-Dithiacyclohexane

When different heteroatoms (viz., N, O, S) are present in the same molecule the alphabetic order is not followed.
The priority order is oxa > thia > aza

e.g., $\overset{8}{CH_3}\text{—}CH_2\text{—}\overset{6}{\text{O}}\text{—}CH_2\text{—}\overset{4}{\text{S}}\text{—}CH_2\text{—}\overset{2}{\text{NH}}\text{—}\overset{1}{CH_3}$

6—oxa—4—thia—2—azaoctane. (*not* 2-Aza-6-oxa-4-thiaoctane in alphabetic order of the heteroatoms). (Rule of lowest sum).

35. [A] Nomenclature of Heterocyclic Compounds

The following heterocyclic compounds have the nomenclature as shown. The priority of getting the lowest locant is of like O > S > N, thus 1,3-oxazole, 1,3-oxathiol are valid. The functional groups, if present, are used as prefixes but the locants are fixed on the heteroatom according to their priority (i.e., O > S > N). The heteroatom is marked.

Examples

[a] Four, five and six membered heterocyclic compounds,

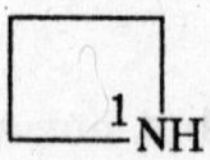

Azetidine

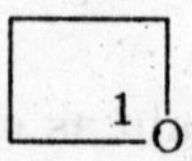

Oxetane

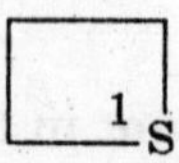

Thitane

Pyrrole

Fulvene

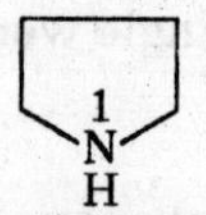

Pyrrolidine

Thiophen

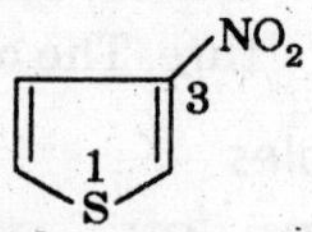

3-Nitrothiophen

Thiolane

Furan

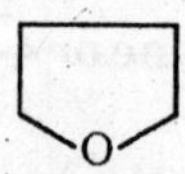

Tetrahydrofuran (THF)

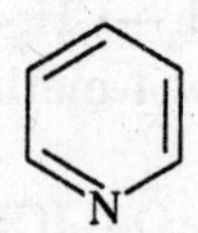

Pyridine

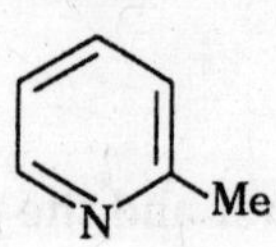

2-Methylpyridine

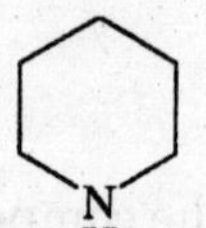

Piperidine

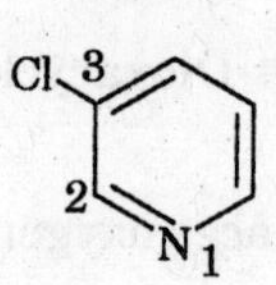

3-Chloropyridine

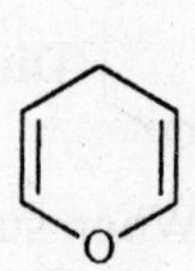

Pyran

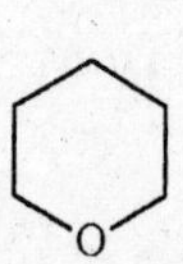

Oxane

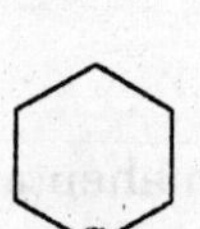

Thiane

Pyrilium cation

Thiopyrilium cation

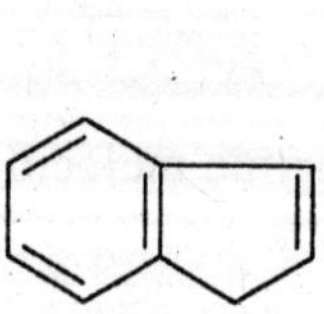

Indene

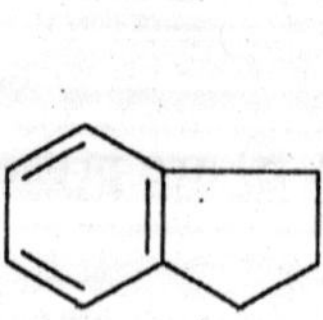

Indane

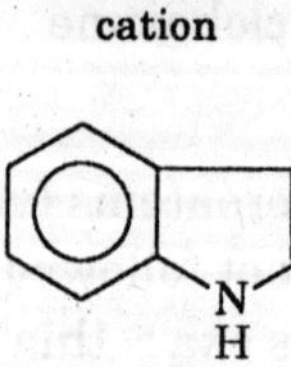

Indole

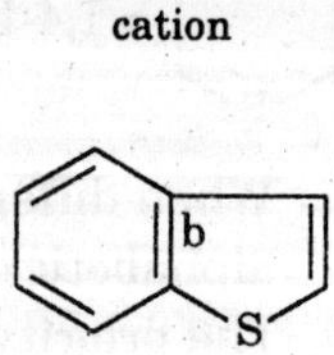

Benzo[b] thiophen

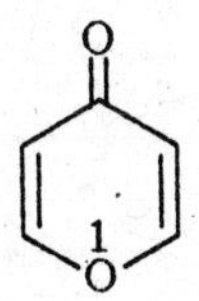

r-Pyrone

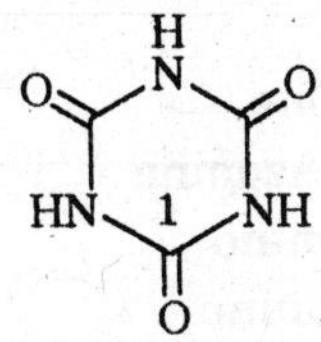

Barbituric acid

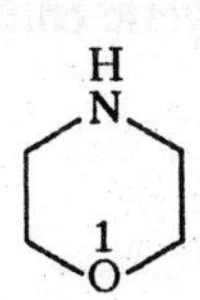

Morpholine

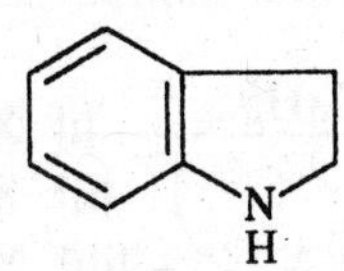

Indoline

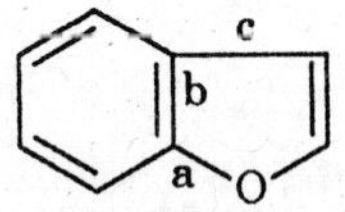

Benzo[b] furan

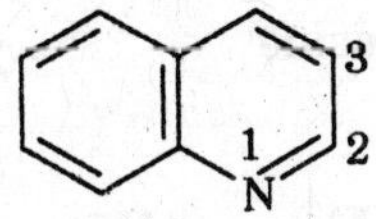

Quinoline

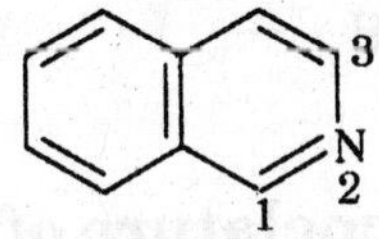

Isoquinoline (Heteroatom *not* 1; exception)

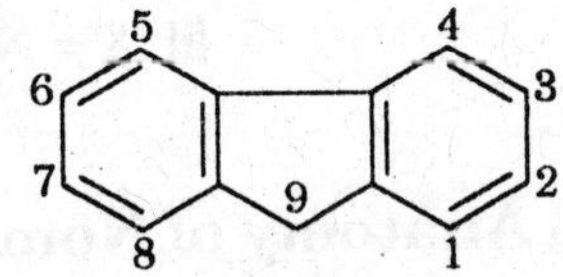

Fluorene

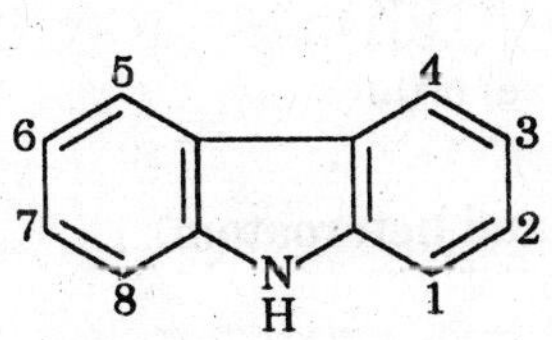

Carbazole

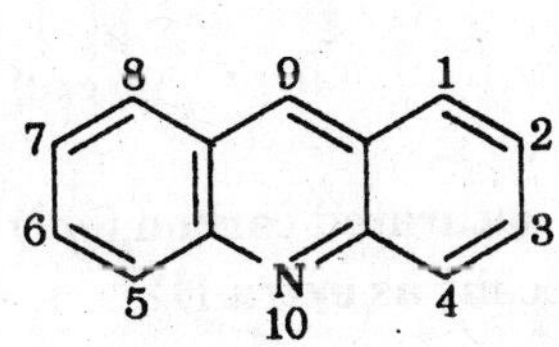

Acridine

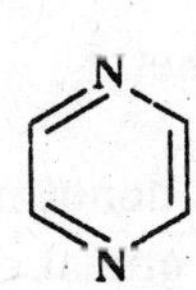

Pyrazine

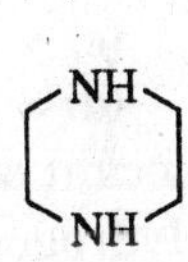

Piperazine

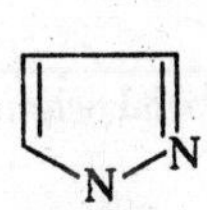

Pyrazole

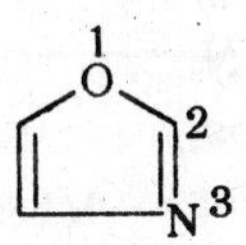

Oxazole

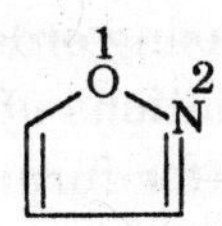

Isoxazole

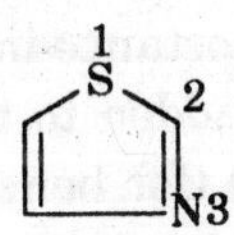

Thiazole

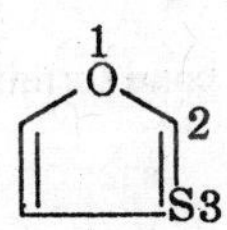

1,3-Oxathiole

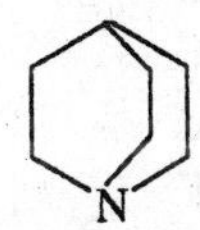

Quinuclidine

[b] Seven membered heterocyclic compounds.

Atom	*Name*
[i] X = NH	[i] 1 H–Azepine
[ii] X = O	[ii] Oxepine
[iii] X = S	[iii] Thiepine

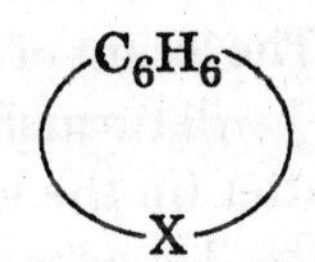

[c] Eight membered heterocyclic compounds.

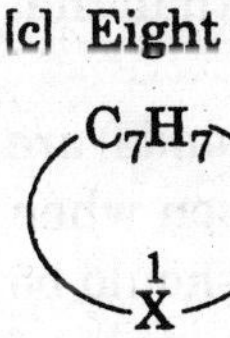

Atom	*Name*
[i] X = NH	[i] Azocine
[ii] X = O	[ii] 2H–Oxocine
[iii] X = S	[iii] 2H–Thiocine

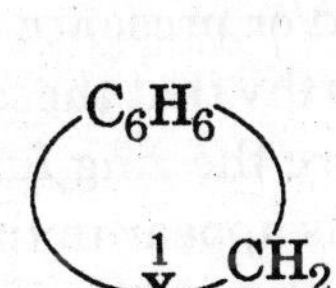

[d] Nine membered heterocyclic compounds.

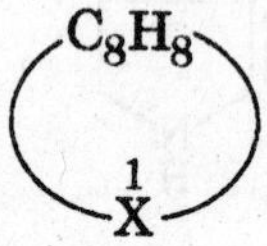

Atom	Name
[i] X = NH	[i] 1H-Azonine
[ii] X = O	[ii] Oxonine
[iii] X = S	[iii] Thionine

[e] Ten membered heterocyclic compounds.

Atom	Name
[i] X = NH	[i] Azecine

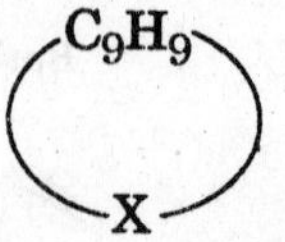

[B] Anatomy of Nomenclature of the Fused Ring Heterocyclic Compounds

The general procedure is shown below.

Legends: (a) = dash present
(b) = dash absent
(c) = square bracket
(d) = comma.

*A*H—Prefix [B, D—E] Suffix
(a) (b)(c)(d) (a) (c)(b)

(e) A = Locant of the extra H denoting the saturated carbon (sp^3) and heteroatom.
(f) H = Hydrogen atom (only one at each locant as extra H).

Prefix : Less important component of the molecule (i.e., smaller in ring-size compared with that of the suffix and/or containing heteroatom of lesser importance in the decreasing order of O>S>N. The 'o' is added to the abbreviations of the prefixes, e.g., benzo (for benzene), furo (for furan), pyrido (for pyridine), pyrano, pyrazolo, etc.

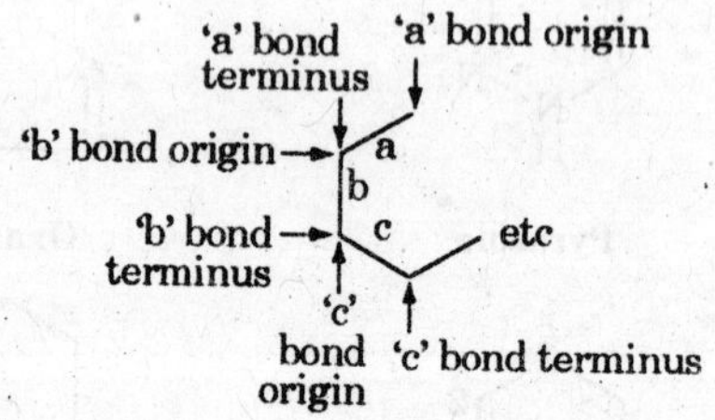

B: Locant of the prefix engaged in the ring fusion between the prefix and the suffix at the 'bond-origin' conforming the direction of travel in the sequence of a → b → c → etc. in the suffix. The origin of any 'bond' (i.e., a, b, c, etc.) will determine B and the 'bond-terminus' will deliver D. The two locants of the prefix are separated by a comma and connected to E by a 'dash'.

D: Second locant of the prefix engaged in ring fusion at the 'bond-terminus'. [It should be noted that if the locant of the bond-origin of suffix be *n* then this bond-terminus will have the locant *n*+1. The locant of prefix connected at *n* will be denoted as B and the other locant connected at *n*+1 will furnish D which will be connected to E by a dash].

E: English letter (in the lower case) marking the bonds in suffix, e.g., bond between 1–2 as (a), 2–3 as (b), 3–4 as (c), etc., which are engaged in ring fusion with the prefix.

Suffix: The parent compound containing N in the ring which is larger in size than that of the prefix and/or presence of other heteroatom(s) like O, S etc.

[It is noteworthy that these locants of the prefix and suffix (to fix the letter of bonds) are considered before the ring fusion. Earlier letters in the alphabet should be chosen when equivalent bonds appear in the suffix. A fresh numbering of the final compound should be given following the rules of lowest number and lowest sum and standard procedure of marking polycyclic compounds containing heteroatoms.]

Examples

[a] Polycyclic aromatic compounds:

[i] Benzo [a] phenanthrene [I] ⟶

(Prefix) (E) (Suffix > Prefix)

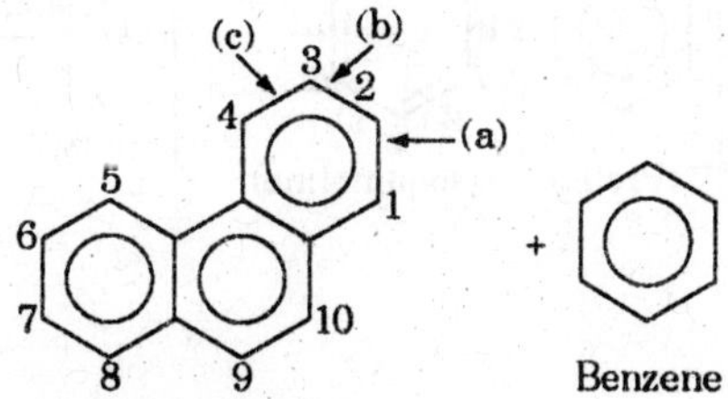

Phenanthrene (marking before bond fusion)

Benzene

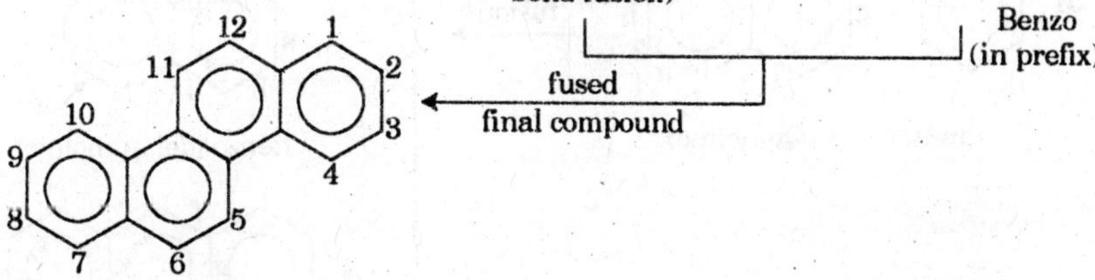

Chrysene [I]

Locants on the final compound (different from those of prefix and suffix).

[b] Benzo [d e f] phenanthrene [II] ⟶

(Prefix)(3 bonds) Suffix (> Prefix)

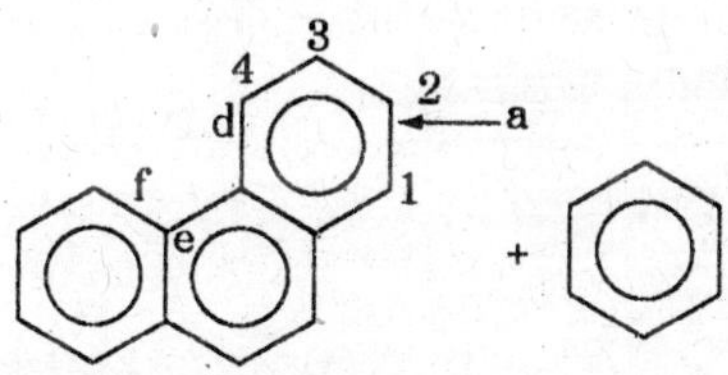

(bonds d, e and f are engaged in ring fusion. The numbering changes in the final compound).

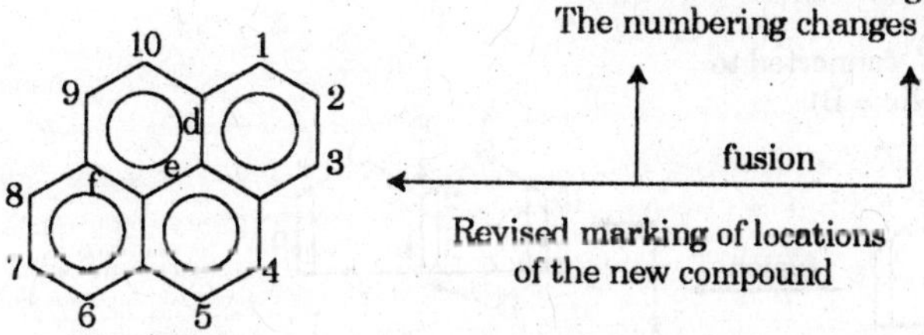

[II] (Pyrene)

[c] Benzo [b] pyridine (quinoline) and benzo [c] pyridine (isoquinoline) vide [e] and [f].

[d] Benzo [b] pyrrole [III] ⟶

Not Benzo [d]-pyrrole (b earlier than d and they are equivalent)

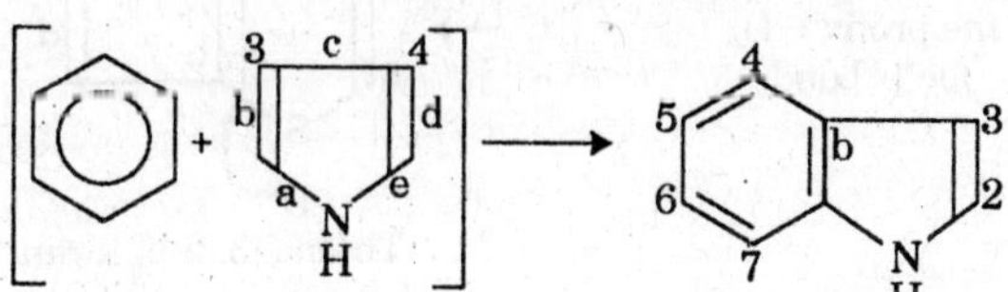

Benzene Pyrrole

[III]

Indole, new marking after ring fusion

[e]

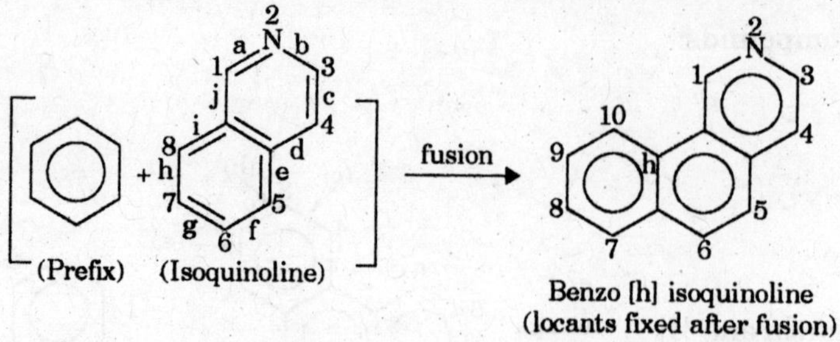
(Prefix)
(Isoquinoline)
fusion
Benzo [h] isoquinoline
(locants fixed after fusion)

[f]

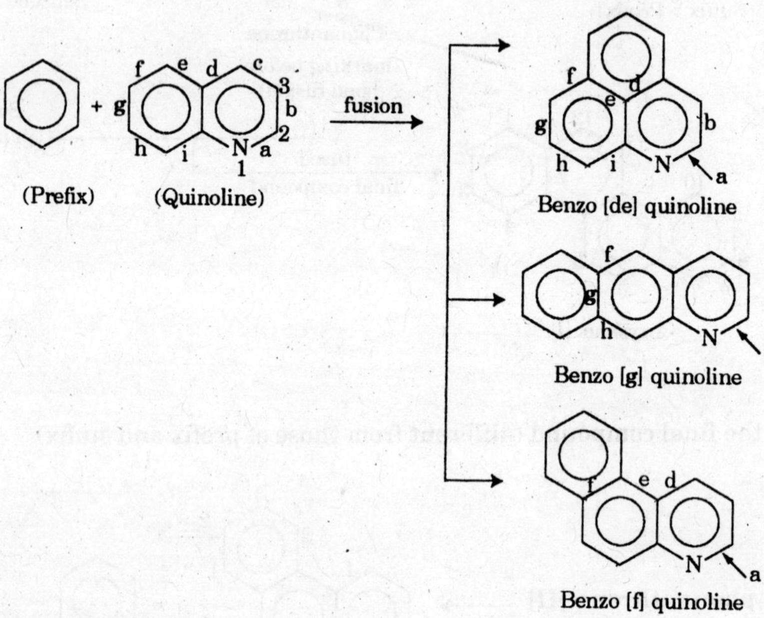
(Prefix)
(Quinoline)
fusion
Benzo [de] quinoline
Benzo [g] quinoline
Benzo [f] quinoline

[g]

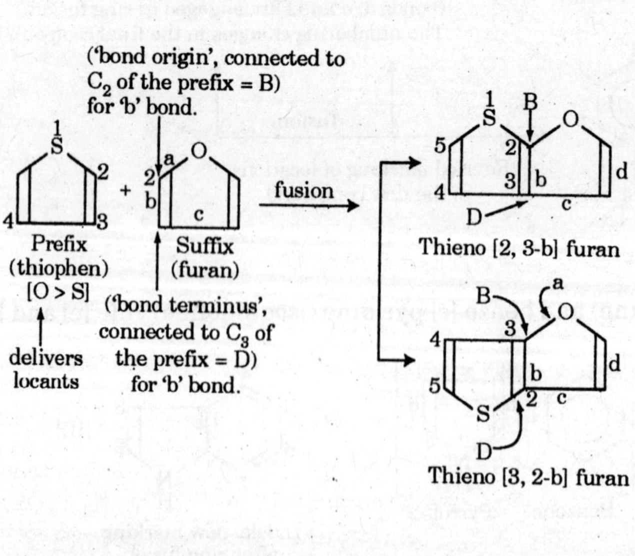
('bond origin', connected to C_2 of the prefix = B) for 'b' bond.
Prefix
(thiophen)
[O > S]
delivers locants
Suffix
(furan)
('bond terminus', connected to C_3 of the prefix = D) for 'b' bond.
fusion
Thieno [2, 3-b] furan
Thieno [3, 2-b] furan

[h] 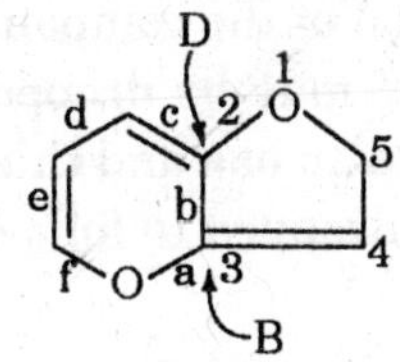 = 2*H*-Furo [3, 2-b] pyran (Numbered on the final compound [A]).

[Pyran > Furan] [B and D—vide Anatomy 4.2] 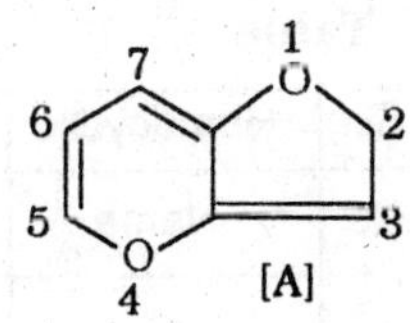

[i] 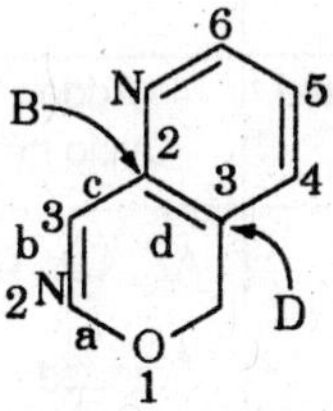 = 5*H*-Pyrido [2,3-d]-*o*-oxazine [O > N]

[B = 'd'-bond origin and D = 'd'-bond terminus].

[j] 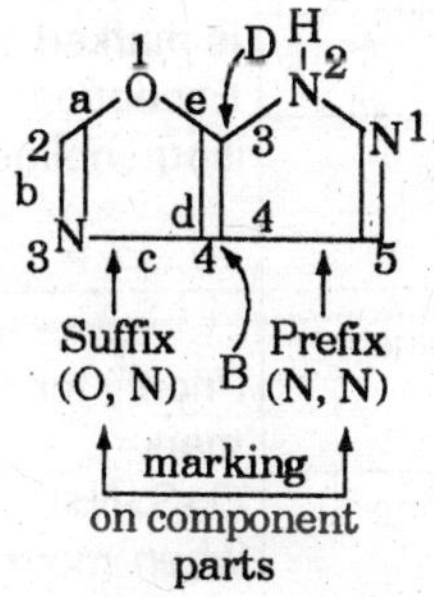 = 1*H*-Pyrazolo [4,3-d] oxazole = 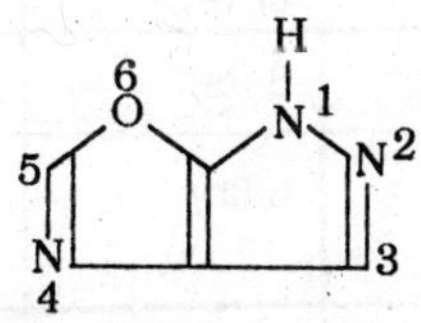

(Final Marking, so 1*H* ital)

[B = d bond origin; D = d bond terminus]

[k] 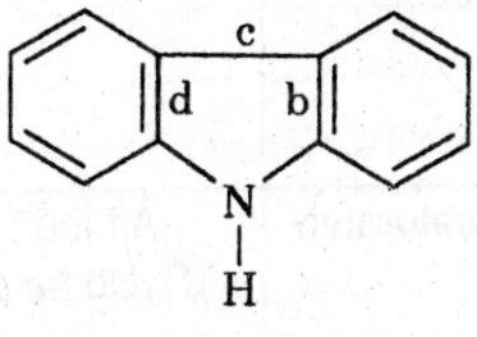 = Dibenzo [b, d] pyrrole (Carbazole)

[l] 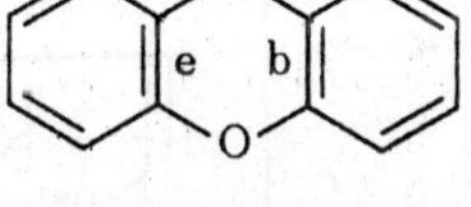 = Dibenzo [b, e] pyran (Xanthene)

30. General Methods of Nomenclature of Heterocyclic Compounds

When the cycle contains one heteroatom then this heteroatom in the monocylic compound should be marked 1, subsequently other functional group, if present, is given the proper locant obeying IUPAC rules applicable to the aliphatic or aromatic compounds.

In the following table the 'ending' related to the size and saturation of the compound is given. These endings are connected to aza (in the table Az+, the 'a' may be dropped, if necessary, to obey the grammar), or oxa (not oxo) and thia (dropping 'a' in oxa and thia). In the table this is shown as Ox+ and Thi+ assuming that 'a' would be dropped to follow the aforementioned grammar.

Table

Sl. No.	Homocyclic Name	Members of the cycle	N—present, Az+ ending (as *)		N—absent, Ox+ ending, (as *) Thi + ending (*)		Remarks
	Means cyclo* as prefix	Root name	Saturated	Unsaturated	Saturated	Unsaturated	Added* = Cyclo before
1.	*propane	3 (ir) cf. t*ri*	*iridine	*irine	*irane	*irene	Az., Ox, Thi, * added
2.	*butane	4 (et) cf. tet	*etidine	*ete	*etane	*ete	Heteroatom is marked 1. Exception isoquinoline.
3.	*pentane	5 (ol)	*olidine	*ole	*olane	*ole	
4.	*hexane	6 (in)	perhydro -a	*ine =a	*ane	*ine	
5.	*heptane	7 (ep) cf. **hep.**	perhydro -b	*epine = b	*epane	*epine	Priority for 1 mark O>S>N>P when more heteroatoms are present in the same cycle.
6.	*octane	8 (oc) cf. **octa**	perhydro -c	*ocine = c	*ocane	*ocine	
7.	*nonane	9 (on) cf. **nona**	perhydro- -d	*onine = d	*onane	*onine	
8.	*decane	10 (ec) cf. **deca**	perhydro -e	*ecine = e	*ecane	*ecine	
Examples	Cyclo* as prefix	Root name	Saturated	Unsaturated	Saturated	Unsaturated	Added* = Cyclo before
1.	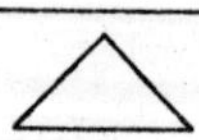 (Cyclo-propane)	(propa)	(ethylene imine) saturated Aziridene	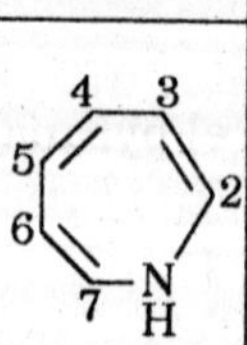 1*H*-Azepine	(ethyliene oxide) 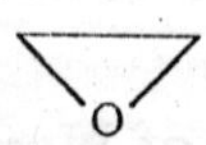 Oxirane		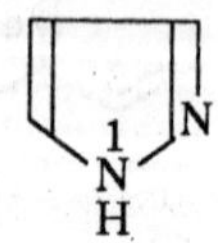 Pyrazole

Contd.

Sl. No.	Homocyclic Name	Members of the cycle	N—present, Az+ ending (as *)		N—absent, Ox+ ending, (as *) Thi + ending (*)		Remarks
Examples	Cyclo* as prefix	Root name	Saturated	Unsaturated	Saturated	Unsaturated	Added* = Cyclo before
2.					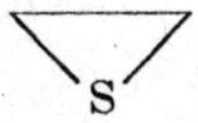 (Thia) + (irane) = Thirane		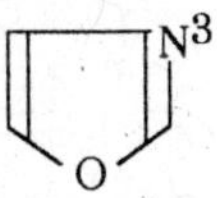 Oxazole
3.				Azecine (10 membered ring)	4-membered  (Saturated) (oxa + etane) = Oxetane	Oxonine (9-membered ring)	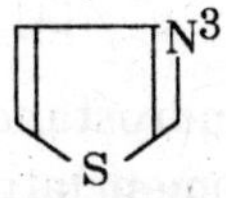 Thiazole 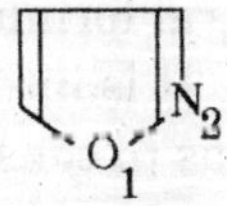 Isoxazole

Examples (miscellaneous)

[i] Ph—CH=N—Ph; 1,2–Diphenylazaethene ('ae' stays)

[ii] $Ph-\overset{1}{C}(=O)-CH_3$; 1-Oxo–1–phenylethane (acetophenone)

[iii] 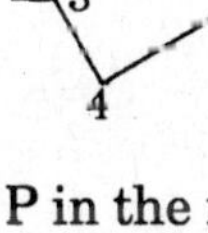 ; 3–Aminopropanoic acid lactam.

[iv] 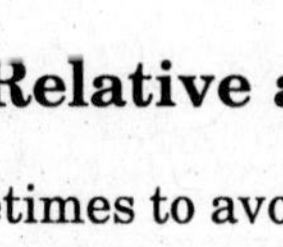 ; 4–Amino–3–methylbutanoic acid lactam.

[v] P in the ring as phospha and Si in the ring as sila started.

37. Relative and Absolute IUPAC Nomenclatures

Sometimes to avoid long and complex IUPAC names of organic compound the name of the parent compound is retained and the substituents are used as prefixes with appropriate locants following all the rules of IUPAC system of nomenclature.

Examples

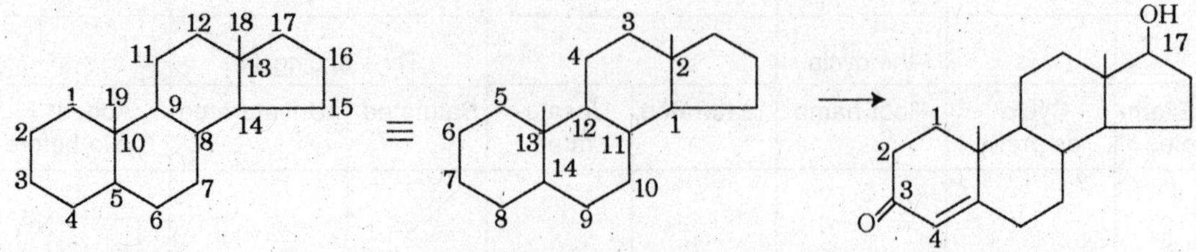

Common name (Androstane) Steroid numbering [A]

Phenanthrene numbering (2,13–Dimethyl–1,2–cyclo-pentanoperhydrophenanthrene) [B] IUPAC name

Common name (Testosterone) [C]

Androstane [A] has the IUPAC name [B] which is long and complex. Hence the IUPAC nomenclature of [C] is done as the derivative of [A]. [C] becomes 17-hydroxy-3-ketoandrost-4-ene which is considered as the relative IUPAC naming because to write down the structural formula of [C] the knowledge of the structure of androstane [A] is essential.

It is interesting to note that a very old method of naming the open chain hydrocarbons was to consider it as a derivative of methane. Thus,

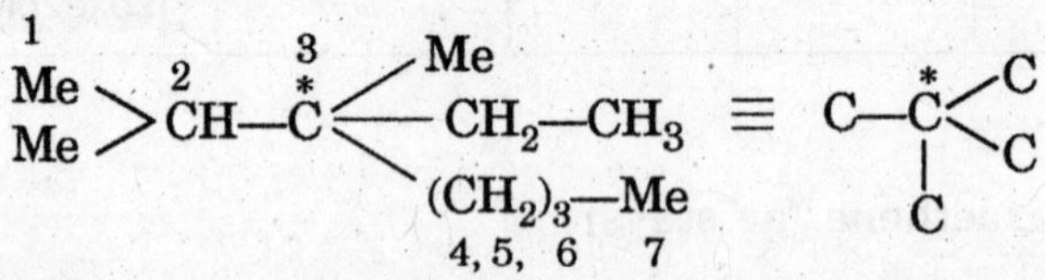

The C* was CH_4, the H atoms have been substituted by four alkyl groups.

The name was butyl ethyl methyl *iso*propyl methane.

The IUPAC name is 3-ethyl-2, 3-dimethylheptane which is absolute naming, because the name is independent of any parent compound.

Other Examples

[i] 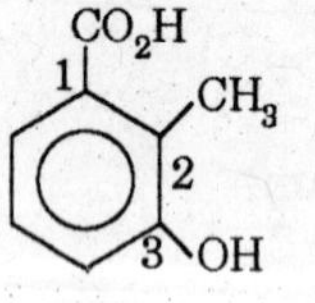

[D]

[ii] 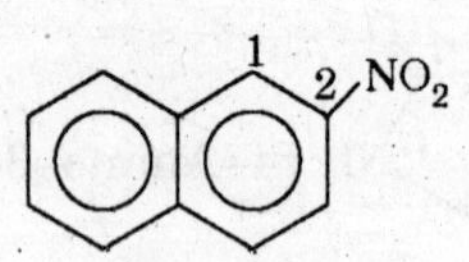

[E]

[iii] 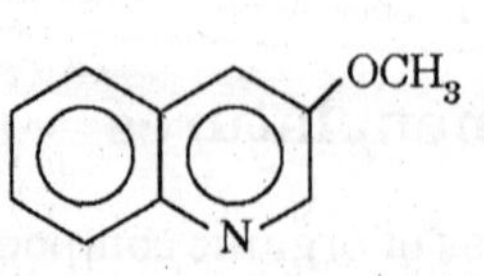

[F]

[iv] 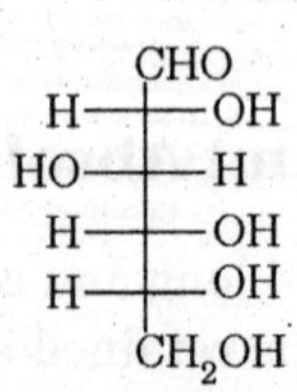

[G]

[i] The name of [D] is 3-hydroxy-2-methylbenzoic acid which may be considered as relative nomenclature because one has to know the structure of benzoic acid as well as the fact that C_1 is the position where $-CO_2H$ group of the compound is fixed on the benzene ring. But the nomenclature of [D] may be also 1-carboxy-3-hydroxy-2-methylcyclohexa-1,3,5-triene which avoids the knowledge of benzene nucleus, and the structure of [D] can be drawn more easily. So the latter naming may be considered as absolute nomenclature of compound [D].

[ii] The relative IUPAC nomenclature of [E] should be 2–nitronaphthalene which is short, easy to remember and simple. But the name is related to the structure of naphthalene which, if unknown, would pose a problem in drawing the structure of [E]. The absolute IUPAC should be bicyclo [4, 4, 0]3–nitrodeca–1(6), 2, 4, 7, 9–pentaene (ae stays). The latter nomenclature is longer than the previous one, but the drawing of the structure of [E] poses no problem.

[iii] The relative IUPAC nomenclature of [F] is 3–methoxyquinoline. But the absolute nomenclature of [F] should be byciclo [4, 4, 0] 4–methoxy–2–azadeca–1(6), 2, 4, 7, 9–pentaene. The advantages and disadvantages of using the two types of nomenclature (i.e., relative and absolute) have been explained before.

[iv] The common name of [G] is glucose (or d-glucose to be more specific). From this name no idea of the structural formula can be derived. But the absolute name of [G] is 2(*R*), 3(*S*), 4(*R*), 5(*R*), 6–pentahydroxyhexanal gives a better exposure of the structure and more chance of drawing the correct structure provided the given 'stereochemical notations' (viz., *R*/*S*) are known beforehand.

38. Stereochemical Symbols, Terms and Expressions in Organic Nomenclature

[a] *Isomers* are compounds having same chemical formula (i.e., molecular formula). These compounds are different from each other due to difference of the arrangements of the constituent atoms in space thus exhibiting different chemical and physical properties (and in some cases optical activity as well as physiological action) of the isomers. This phenomenon is known as *isomerism*.
All isomers will have same molecular weight but all compounds having same molecular weights may not be isomers.
Stereochemical characters like *R*/*S*, *E*/*Z*, *cis*/*trans*, *erythro*/*threo* ... etc., are underlined in hand written copies or italicised when printed.

Examples

[i] $CaCO_3$ (inorganic) and CrO_3 (inorganic) are not isomers though each one of these have molecular weight 100, and they are not organic.

[ii] $CaCO_3$ (inorganic, mol. wt. 100) is not isomeric with C_7H_{16} (organic, mol. wt. 100).

[iii] C_7H_{16} (organic; mol. wt. 100) is not isomeric with $C_5H_{12}N_2$ (organic; mol. wt. 100).

[iv] C_7H_{16} has many isomers which differ in structural formula. Thus ,

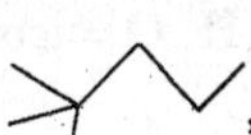, 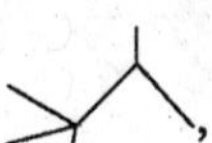, 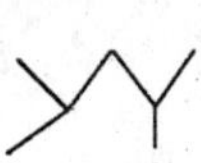 and many more compounds having molecular formula C_7H_{16} (so molecular weight 100) are isomeric.

Different Types of Isomerism

The following chart is illustrative.

Isomerism

- **Structural isomerism**
  - [a] Chain isomerism [C]
  - [b] Positional isomerism [P]
  - [c] Functional group isomerism [F]
- **Stereo isomerism** — [C], [P], [F] same but the spatial arrangement of the atoms are different
  - Optical isomerism [Opt.]
  - Geometrical isomerism [Geo.]

Examples

[a] [C]: Skeletal structure is different. Sometimes called *nuclear* isomerism.

[i] $C_4H_{10} \equiv Me(CH_2)_2—CH_3$ and $Me_2CH—CH_3$, [ii] C_5H_{12} =  , 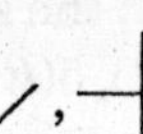 ,  ,

[b] [P]: Skeletal structure is same, functional group is same but the position of functional group is different (i.e., locants are different).

[i] $C_4H_{10}O$ or C_4H_9OH = alcohol: 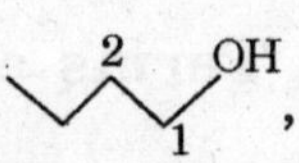 , 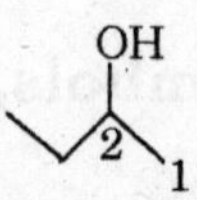

(butan-1-ol) (butan-2-ol)

[ii] 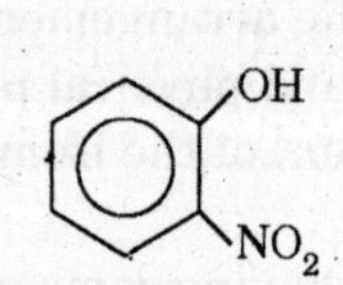 , 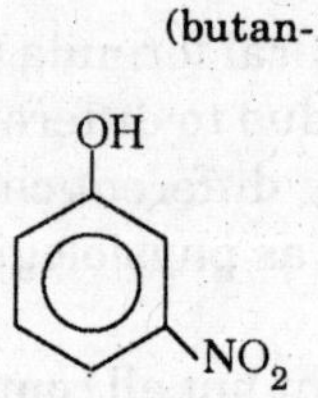

(ortho) ***(meta)***

[c] [F]: Skeletal structure may be same or different but functional group is different.

[i] $CH_3—CH_2—OH$, $CH_3—O—CH_3$; skeletal structure is different, functional group is different.

[ii] $CH_3—CO—CH_3$, $CH_3—CH_2—CHO$; skeletal same, but functional group is different.

[iii] $CH_3—CO—OCH_3$, $CH_3—CH_2—COOH$; same as example [i] above.

[d] Isomerism arising from combinations of [C], [P] and [F] are also possible.

[i] 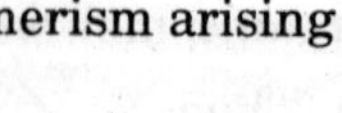 , 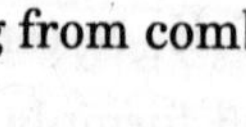 , 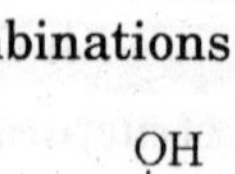 , 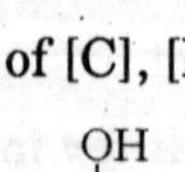 , 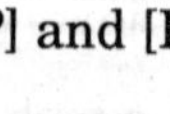 , 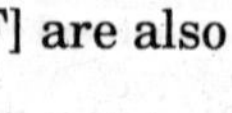 **= $C_4H_{10}O$ = isomers.**

[A] [B] [G] [D] [E] [F]

[A] and [D] = [C] + [P]; [A] and [E] = [C] + [F]; [E] and [F] = [C]; [A] and [G] = [P]

[ii] *Tautomerism*: Same compound existing in two different functional group isomerism due to rapid exchange of an atom or group from one locant to another in an equilibrating condition. Thus,

$$\underset{\text{(keto form)}}{CH_3-\underset{\underset{O}{\|}}{C}-CH_2-CO_2C_2H_5} \rightleftharpoons \underset{\text{(enol form i.e., ene+ol)}}{CH_3-\underset{\underset{OH}{|}}{C}=CH-CO_2C_2H_5} = \text{Tautomers} = C_6H_{10}O_3$$

This phenomenon is known as 'Keto-enol tautomerism' of ethyl acetoacetate.

39. Optical Isomerism

Optically active (or non-active) isomers related to the influence of plane polarised light on the compound are of three types.

[i] That molecule which turns the plane polarised light in the right-hand side of the observer (i.e., clockwise rotation) is termed dextrorotatory (i.e., *d* or +)

[ii] That molecule which turns the plane-polarised light in the left direction (i.e., anticlockwise rotation) is termed levorotatory (i.e., *l* or –).

[iii] That molecule which has half portion *d* and the other half *l* cannot turn the plane polarised light to any direction, thus rotation is zero degree. This molecule is termed *meso* isomer.

[iv] It is to be noted that the (+) or (–) signs do not represent the electric charge or the mathematical plus or minus, but they represent the direction of rotation of plane polarised light caused by the compound under investigation.

[v] It is to be further noted that an equimolar mixture of *dextro* and *levo-isomers of the same* compound will exhibit zero degree rotation (i.e., optical inactivity) just like the *meso*-isomers. This mixture is termed *racemic* compound (*dl* or ±). The *racemic* compound may be separated into its components (i.e., *d* and *l* separated), because it is made of two molecules, but the *meso*-compound is a single molecule, hence there is no question of separation (called resolution) of it into *d* and *l* isomers.

[vi] The stereochemical notations and symbols are generally underlined (or italicised, if printed).

[vii] Optically active compounds are dissymmetric.

Examples

[a] Isomerism of tartaric acid, $C_4H_6O_6 \equiv \begin{array}{l} {}^*CH(OH)-CO_2H \\ \;| \\ {}^*CH(OH)-CO_2H \end{array}$ (C* = asymmetric carbon).

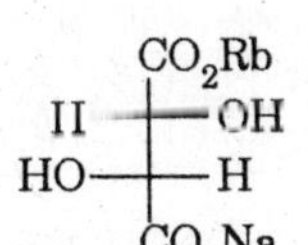

[R]

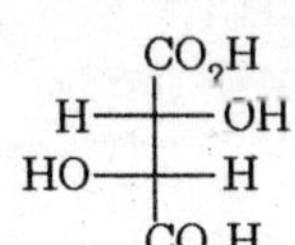

[A]

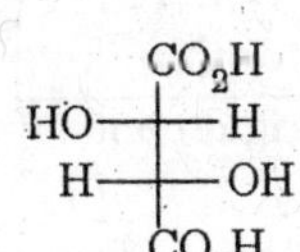

[B]

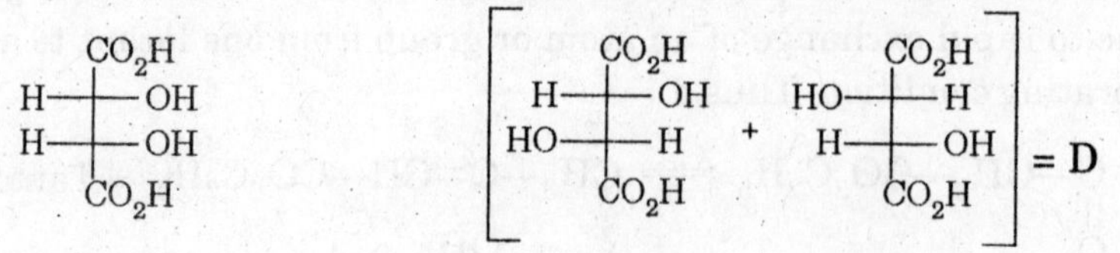

[C] [A] + [B] (1:1, w/w)

[R] is the absolute configuration of rubidium sodium salt of (+) tartaric acid as determined by zirconium K_α – X-ray. So (+) tartaric acid (i.e., *d* or dextrorotatory tartaric acid) must have stereostructure [A]. The mirror image of [A] is (–) tartaric acid (i.e., *l* or levorotatory tartaric acid) must have stereostructure [B]. The *d* and *l* isomers are non-superimposable stereoisomers and called enantiomers of each other. The stereoisomers which are not enantiomers are called diastereomers (formerly called diastereoisomers). Thus optically inactive *meso* tartaric acid [C] is the diastereomer of each of *d*– and *l*– tartaric acid. The 1:1 (w/w) mixture of *d* and *l* stereoisomers is also optically inactive as the +x° rotation of *d* is cancelled by –x° rotation of *l*. This is called the *racemic mixture* of tartaric acid [D], and not an isomer of *d* or *l* (and also *meso*), because its formula weight will be twice of any one of the stereoisomers. So tartaric acid has four *varieties* (not four stereoisomers) like [A], [B], [C] and [D] and three stereoisomers [A], [B] and [C].

40. [A] Geometrical Isomerism

This isomerism arises from molecules having restriction of rotation of bond(s) due to C=C or a cycle in the molecule. Formerly this isomerism was also known as *cis/trans* isomerism. The atom or group attached to the carbon at the 'rigid bond' may be classified as

[i] Same ligands, e.g., A/A; 1/1; CH_3/CH_3; OH/OH, etc., make pair.

[ii] Similar ligands, e.g., A/B; 1/2; CH_3/C_2H_5; R/R_1; Cl/Br; NH_2/OH, etc. make pair.

[iii] Rest or remaining ligands, e.g., ligands which are *not* same or similar, make pair.

When same pair is situated on the same side of the *rigid* bond then *cis* isomer is formed. The *cis/trans* are to be underlined, they are called configurational notations (i.e. related to the constitutional aspect) of the compound. The *cis/trans* are diastereomers (*not* enantiomers). Thus,

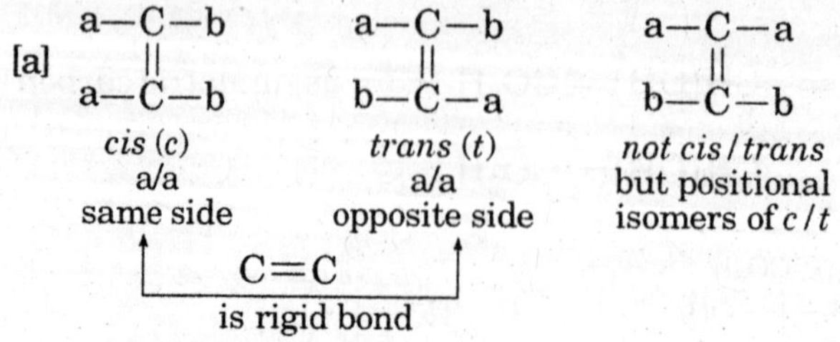

Therefore, three isomers of $C_2a_2b_2$ are possible. Thus three isomers of $C_2H_2\ (CO_2H)_2$ are possible, viz., x, y, and z.

These examples are of same/same ligands type.

[b] 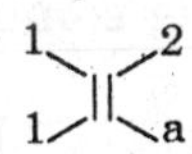 , 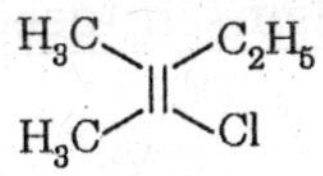 , 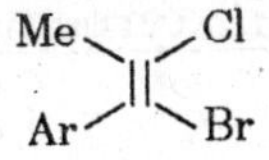 , 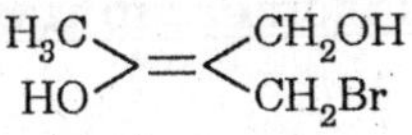 , 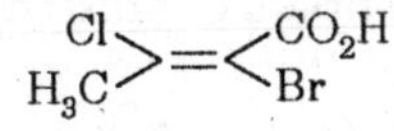

cis	*cis*	*cis*	*trans*	*trans*
(same/rest)	(same/rest)	(similar Me/Ar)	(similar OH/CH_2OH rest CH_3/CH_2Br)	(similar halogens/ rest CH_3/CO_2H)

The examples of [a] and [b] are arising from the restricted rotation of the C=C bonds. If this bond is converted into σ bond then the *cis/trans* isomersims do not arise. Another type of restriction of rotation of σ bond is exhibited in cyclic compounds. The *cis/trans* isomers are generally expressed through three types of bond drawing.

[i] This is β (*pr.* beta) bond, which propagates towards the observer (i.e., above the plane of the paper).

[ii] - - - - - - - - *or* |||||||||||||||| This is α (*pr.* alpha) bond, which goes towards the opposite side of the observer (i.e., down the plane of the paper).

[iii] ——, This is *on plane* bond, which lies *on the* plane of the paper.

[B] Geometrical Isomerism in Alicyclic Compounds

The cyclic compounds possessing α and β substitutes may generate *cis/trans* isomers (i.e., diastereomers) which may be optically active compounds depending on the position of substitution, specially in the six membered ring.

[i] α, α or β, β bonds are in the same direction hence they will generate *cis*-isomers,
[ii] α, β bonds are in opposite direction, hence they will generate *trans*-isomers.

Examples

[A] [a] 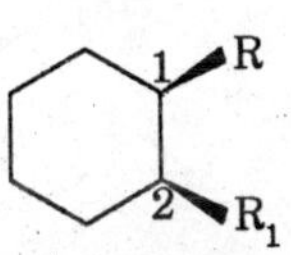

R = R_1 or R ≠ R_1
R (and R_1) = OH, Me, Hal, etc.

β, β substitution at C_1, C_2
cis (optically inactive, if R = R_1)
(optically active, if R ≠ R_1)

[b] 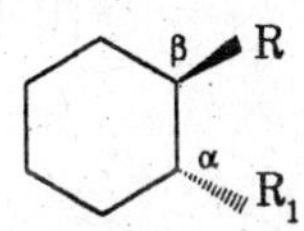

α, β substituents, at C_1, C_2
trans (optically active)

[c] 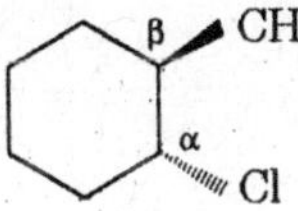

(α, β) *trans*
(optically active)

[d] 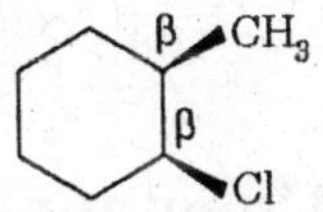 same as 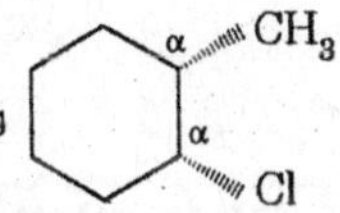

β, β-*cis* ≡ α, α-*cis* (optically active)

Examples of *cis-trans* isomerism (i.e., diastereomerism) in fused ring compounds.

[i]

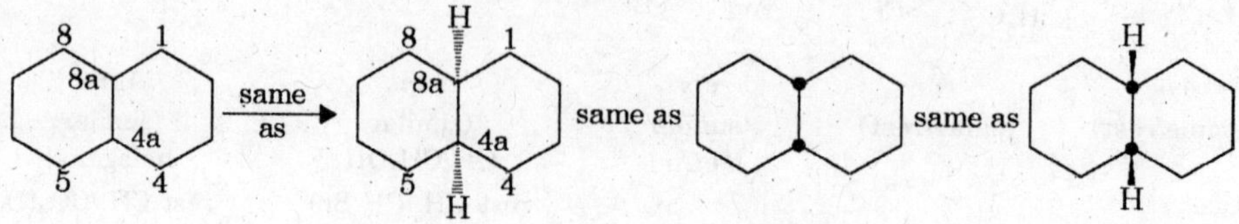

4a and 8a are ring junctions (i.e., bridge-head) | *cis*-(α, α) | *cis*-(β, β) ● means βH | *cis*-(β, β) decalin

[ii]

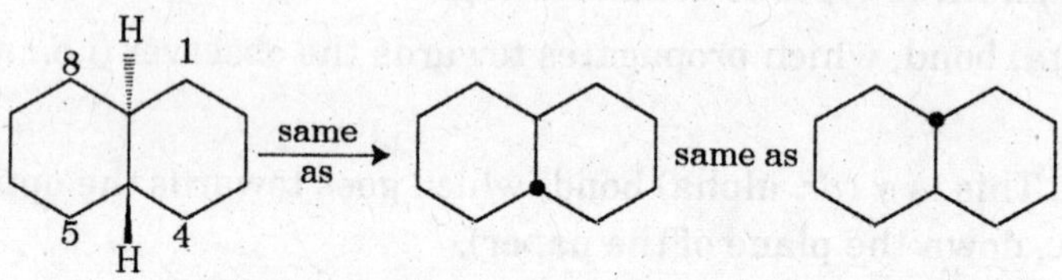

● (dot) mark is β

(α, β) *trans*-decalin | (α, β)-*trans* | (β, α)-*trans* decalin | (not dotted is α)

4a, 8a are ring junctions

[iii]

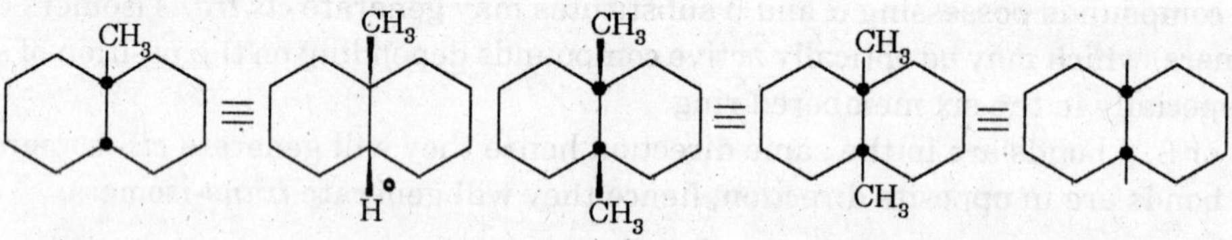

(vide short-hand structure)

(β, β)-*cis* | (β, β) *cis* | cis-(β, β) | *cis* | *cis*

as well as

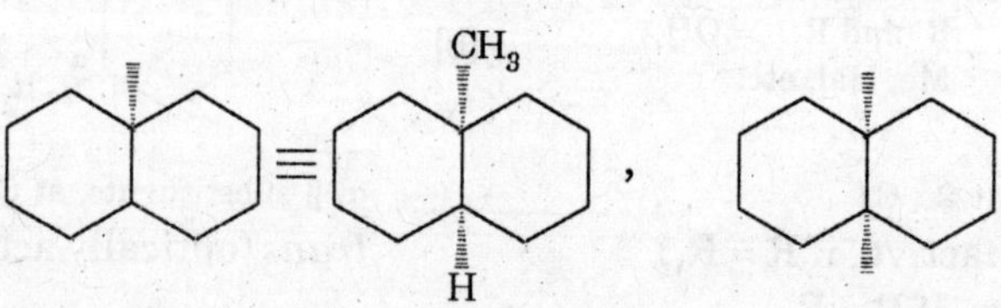

α,α-*cis* | α,α-*cis* | (α, α) *cis*

[C] Connected Ring Compounds

These alicyclic compounds are not of the fused ring class. Two rings at a time is considered to be connected by a single bond. The stereochemical terms are *syn* (cf. *cis*) and *anti* (cf. *trans*) as shown below. The rings are marked as A, B, C, ..., etc. [● is Hβ].

[a] [i] 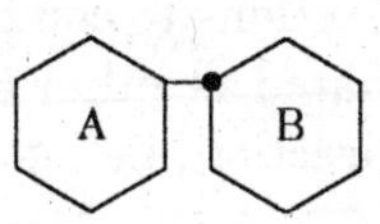

[ii] 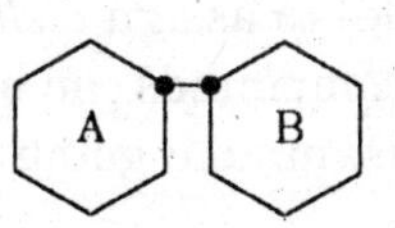

same as 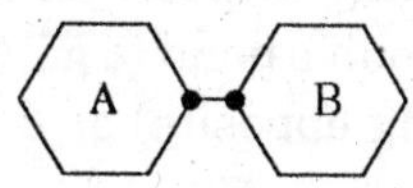

and 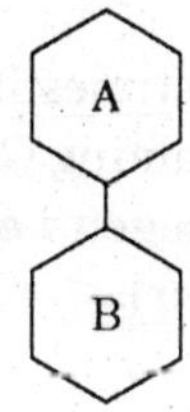

(α, β) *anti* (β, β), *syn* *syn* *syn*

[b] [i] 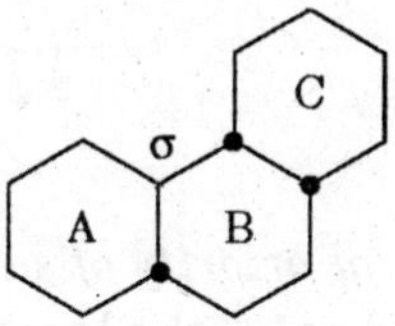

A/B = *trans* (β, α) = *t*
A/C = σ bond connection,
= *anti* (α, β) = *a*
B/C = *cis* (β, β) = *c*
This is *t* – *a* – *c* or *trans-anti-cis* isomer of perhydrophenanthrene (PHP)

[ii] 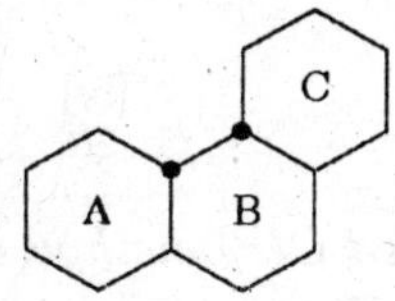

A/B = *t*
A/C = *syn* (i.e., *s*)
B/C = *t*
This is *t* – *s* – *t* perhydrophenanthrene (PHP)

[iii] 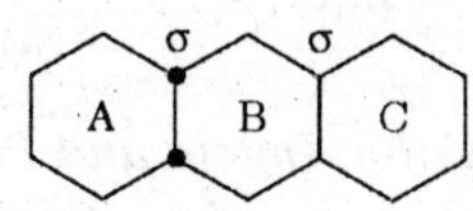

A/B = *cis* = *c*
A/C = *anti* = *a*
B/C = *cis* = *c*
This is *c* – *a* – *c* perhydroanthracene (PHA)

[iv] 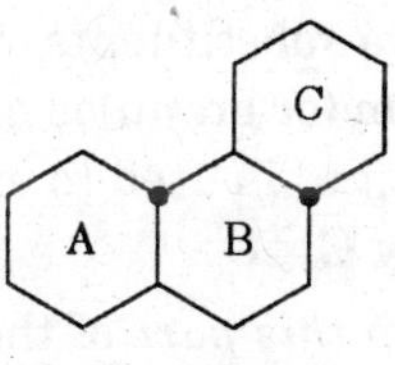

t – *a* – *t* (PHP)

[v] 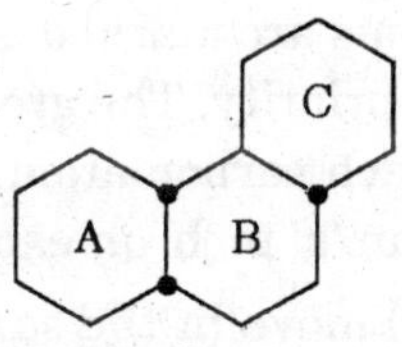

c – *a* – *t*
≡ *t* – *a* – *c*
(PHP)

(note [b] [i])

[vi] 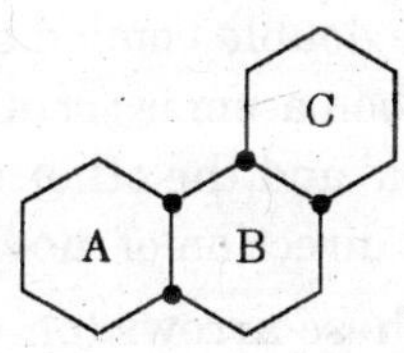

c – *s* – *c* (PHP)

[vii] 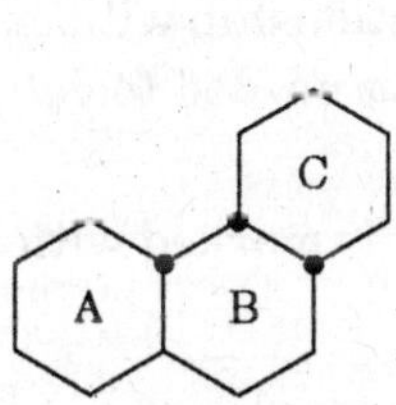

t – *s* – *c* (PHP)

41. *E/Z*-notations at Double Bonds

Sometimes it becomes difficult (or impossible) to assign *cis/trans* notations to compounds containing C=C double bond (e.g., A). Such compounds can be designated as either *E* (from Entgegen i.e., going opposite) or *Z* (from zusammen; going same direction, i.e., going together).

[A] [i] 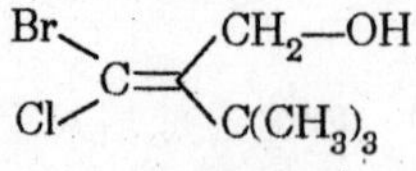 [ii] 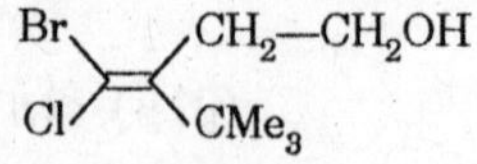

[iii] 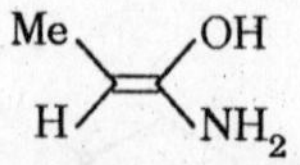 [iv] 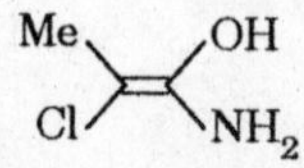

[a] *Cahn, Ingold and Prelog (called CIP) method of determination of priority of substitution on carbons (vide Sec. 52.)*: The maximum priority is marked as 'a', the least priority is 'd' and the priority decreases in the order a > b > c > d (in a few examples the order is presented as 1 > 2 > 3 > 4). This priority is determined by 'sequence' rules as shown below.

[i] The atoms directly attached to the carbon (under consideration) are compared. The atom which has the *greater atomic number* gets priority over the other atoms on the same carbon. Thus atomic number decreases as a > b > c > d.

[ii] If the atoms directly attached to carbon are same, then the next atom is similarly compared with each other to determine the priority.

[iii] Other three rules have been given elaborately under *R/S* notations (vide Sec 52)

[b] The double bonded carbon atoms are marked as $C_1{=}C_2$. The two substituents on each carbon atom is scrutinised for priority. The greater priority atom (or group) is marked as 'a' and the other as 'b' on each carbon atom. The → (arrow sign) is used to express the direction of movement from 'a' to 'b' on each carbon holding C=C.

[c] If these arrows (on C_1 and C_2) move in the *same direction* then *this part* of the compound is marked as *Z*. When the movement is in the *opposite direction*, then this part of the molecule (holding C_1 and C_2 double bond) is termed E. If the molecule holds only one C=C then the whole molecule will be either *Z* or *E*.

[d] The notation *E/Z* has greater application compared with *cis/trans*. The *E* compound may be *cis* or *trans*, such is the case with *Z* (i.e., *E/Z* has no relationship with *cis/trans* notations). The 'c' and 'd' in the priority list a > b > c > d are not needed to determine notations.

[e] If the molecule holds more than one C=C units, then each unit is marked with *E/Z* to designate the whole molecule.

Examples

[i] With only one C=C [A (i), (ii), (iii) and (iv) shown before.]
The compound A[i] is *Z*; A[ii] is *E*; A[iii] is *Z*; A[iv] is *E* configurations.

Explanation

A.[i] 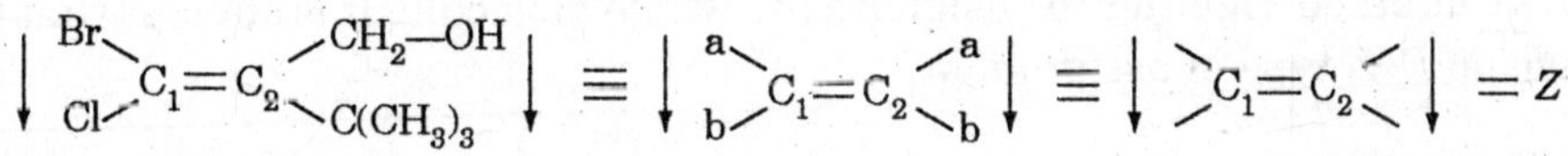

Priority Br > Cl at C_1 or Br = a ; Cl = b at C_1
Priority C (H, H, O) > C (C,C, C) at C_2
or CH_2OH = a, $C(CH_3)_3$ = b

a → b at C_1
a → b at C_2
(same direction)

So the compound A[i] has Z configuration (The compound A[i] is designated as Z)

[**Note**: one atom of greater atomic number has priority than more atoms of lower atomic number, so the atomic numbers are not added for comparison of priority]

A.[ii] 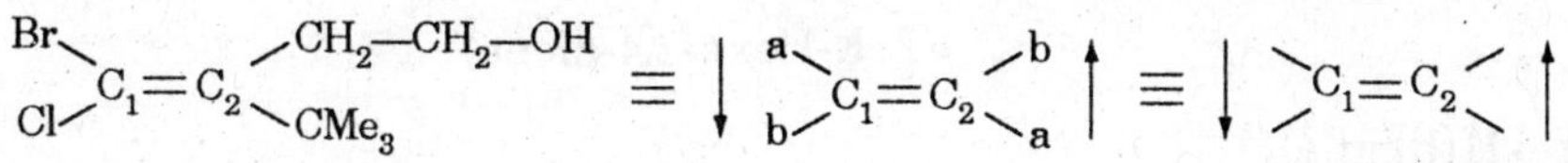

So this A[ii] is of *E*-configuration

Priority, Br (a) > Cl (b) at C_1
C (C, C, C) (a) > C (H, H, C) (b) at C_2
≡
a → b at C_1
a → b at C_2 (Opposite direction)

A.[iii] 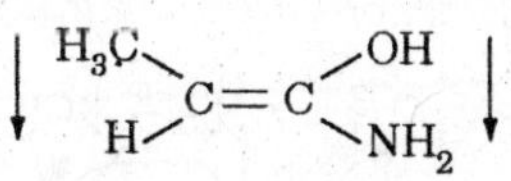

(C Vs. H) (O Vs. N)
'*Z*'

A.[iv] 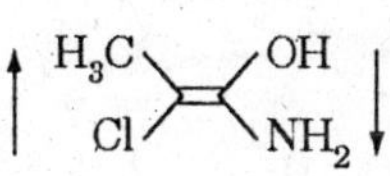

'*E*'

A.[v] 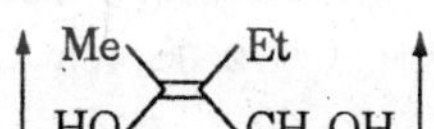

cis (also *Z*)

A.[vi] 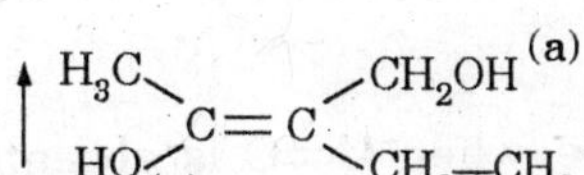

(O Vs. C); (O, H, H Vs. C, H, H)
CH_3 is similar to CH_2–CH_3 (so *trans*)
trans (also *E*)

A.[vii] 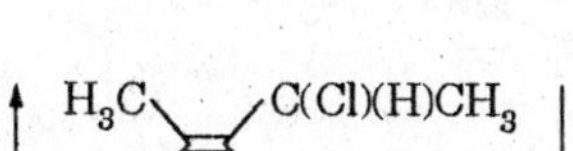

(O Vs. C); (Cl, H, C Vs. H, H, O)
OH is similar to CH_2OH (*cis*)
cis (*E*)

A [viii] 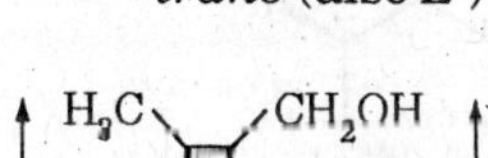

(HO and CH_2–OH similar)
trans (*Z*)

[**Note**: *cis* and *trans* have no relationship with *E*/*Z* configurational notations.]

[f] When more than one C=C are present then each C=C is designated by E/Z, finally these are combined together to designate the whole molecule. It is obvious that $C{=}CA_2$ will have no E/Z type of assignment.

Examples

[i] 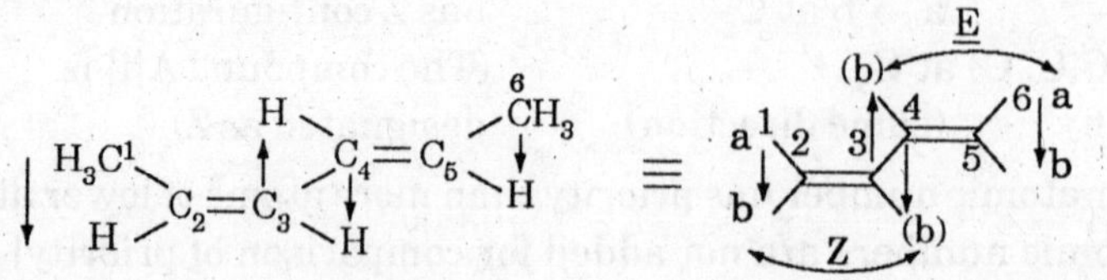

(Z, E-Hexa-2,4-diene)

$$\left.\begin{array}{l} CH_3\,(a)\rightarrow H(b) \text{ at } C_2 \\ C_4\,(a)\rightarrow C_3H \text{ at } C_3 \end{array}\right\} Z,$$

$$\left.\begin{array}{l} C_3\,(a)\rightarrow C_4H \text{ at } C_4 \\ C_6\,(a)\rightarrow C_5H \text{ at } C_5 \end{array}\right\} E$$

[ii] 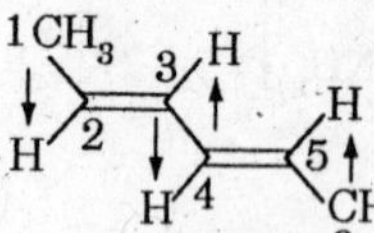

E, *E*

(*E*,*E*)-Hexa-2,4-diene

[iii] 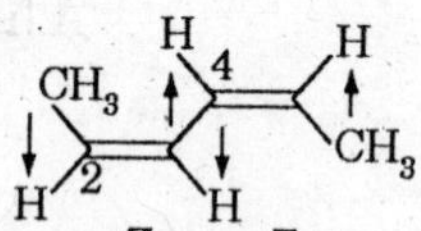

Z, *Z*

(*Z*,*Z*)-Hexa-2,4-diene

[iv] 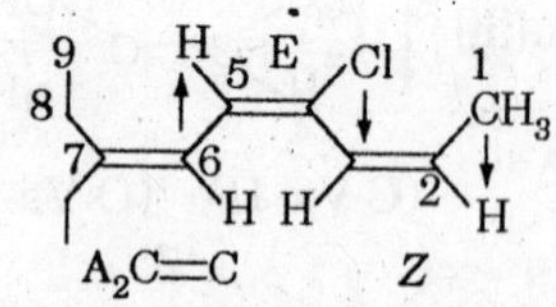

(*Z*,*E*)-compound (not E, Z)

2, 3 > 4, 5

[**Note**: when C=C is absent, there is no E/Z configurational notation but *cis*/*trans* notations may exist with or without optical activity).

Examples

[i] 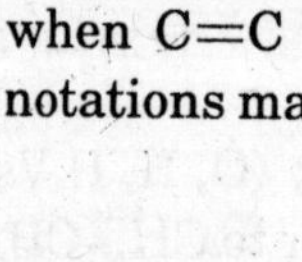

trans; optcially active; no E/Z

[ii] 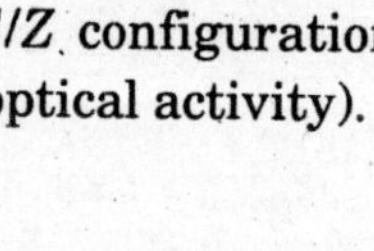

cis; optically inactive; no E/Z

42. S-*cis* and S-*trans* Notations

When two C=C are connected by a single bond, and the rotation of the single bond is restricted then the S-*cis*/S-*trans* configurations are assigned to the compounds. When these C=C are on the same side of the single bond then this will be designated as S-*cis* while the opposite orientation of the double bond will be described as S-*trans* (the S stands for the single bond holding the two C=C). In the following example the single bond is shown by the arrow.

Examples

[i]

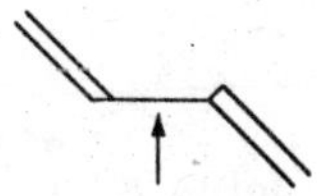

S-*trans*
Two C=C are in opposite direction

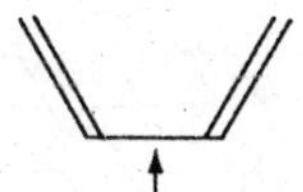

S-*cis*
Two C=C are in same direction

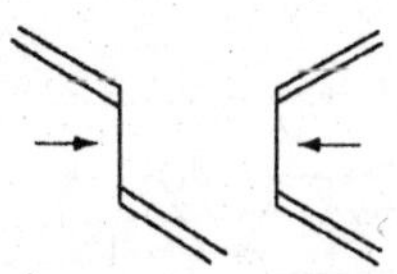

S-*trans* S-*cis*

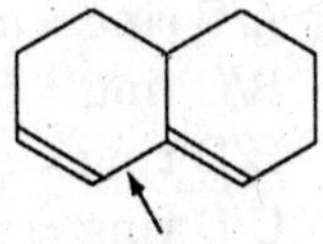

S-*trans*

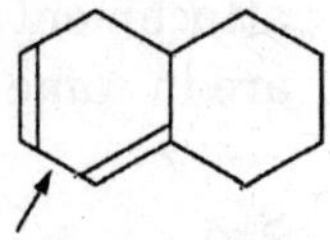

S-*cis*

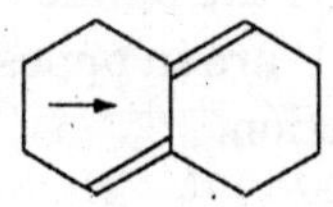

S-*trans*

43. *Syn/Anti* Notations

In the case of oximes of carbonyl compounds the group on carbon (attached by the double bond to the N holding the OH group) when situated on the same side of the OH group will furnish the *syn* derivative. When a group is oppositely situated to the OH group the derivative will be *anti* (*this is configuration here*).

Examples

[i] 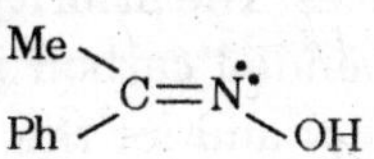

Naming: *Syn*–phenylacetophenone oxime (OH/Ph same side)
or
Anti–methylacetophenone oxime (OH/Me opposite)

[ii] 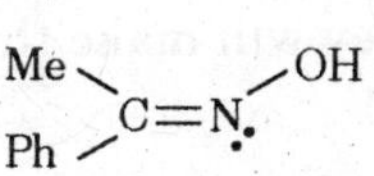

Naming: *Syn*–methylacetophenone oxime (OH/Me same side)
or
Anti–phenylacetophenone oxime (OH/Ph opposite)

[iii] **During conformational notation**: (vide infra) Similar or same group (or atoms) at 180° direction with each other will furnish the *anti conformation*.

Examples

[a] 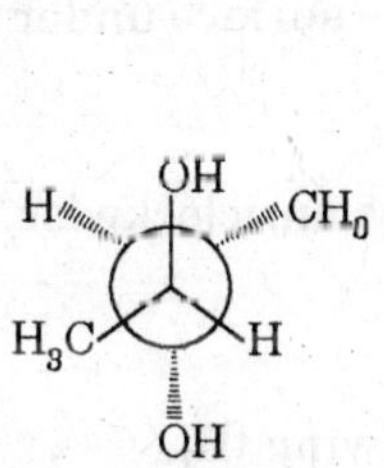

Anti-conformation (old notation)
Staggered conformation (current notation)

[b] 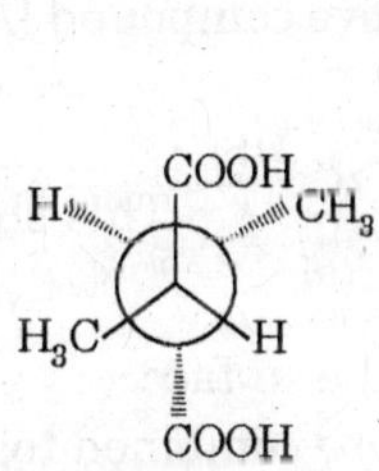

Anti (old convention)
Staggered (current conformation)

[iv] Two alicyclic rings are connected by a single bond. The relative position of these connecting points are expressed as *syn/anti* conformations.

Examples

[a] 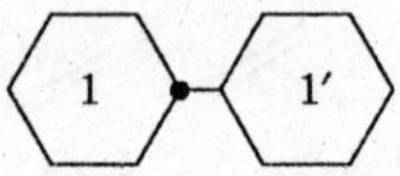

The connecting points that is the ligands at 1 and 1′ are in opposite direction.
Anti–

[b] 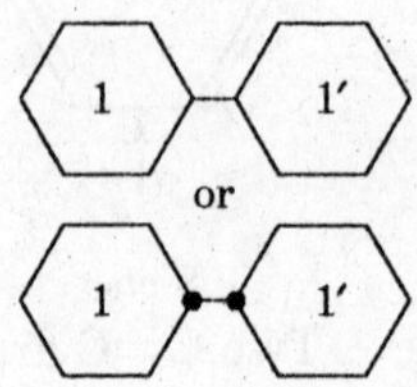

The two points of attachment of ligands are in same direction

Syn

[c] 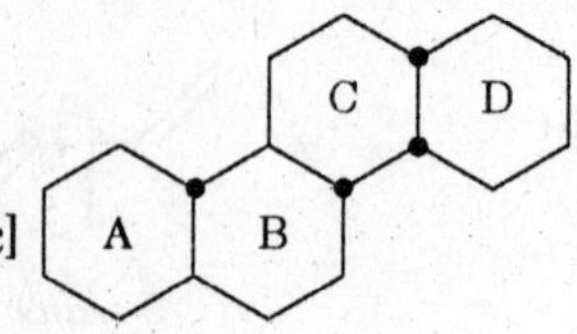

A/B ring = *trans*,
A/C ring = *anti*,
B/C ring = *trans*,
B/D ring = *syn*,
C/D ring = *cis*.
Designation:
trans–anti–trans–syn–cis
(i.e., *t–a–t–s–c*)

44. *si*/*re* Notations at Compounds with Double Bonds

The sp^2 hybridised carbons are planar (as in C=C, C=O, etc). So the two sides of the plane (called faces) are of different orientation with respect to the substituents. The CIP sequence rule is applied (vide infra) to determine the character of the faces. The priority of each one of the ligands of the double-bonded carbon is determined and marked as (a), (b) and (c) and let the order of priority be a > b > c. If the journey from a → b → c be clockwise, the face will be *re* (from *R*) and the anticlockwise journey will make the face *si* (from *S*). Let the taken compound (b) (c)C=(a), marked [A].

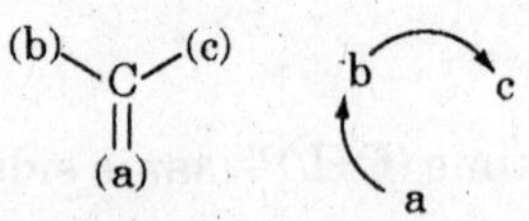

[A]

Examples

[i] This is clockwise when seen from the top surface of the paper. So this face of the compound is *re*-face.

[ii] When the above compound [A] is observed from the surface under the paper, this will look like 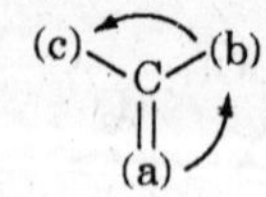 . This shows (a) → (b) → (c) as anticlockwise. So this face of the compound [A] is *si* face.

The [i] and [ii] are combined together to give the drawing [B].

This face is *si* (down the plane of the paper (a) → (b) → (c) anticlockwise movement.

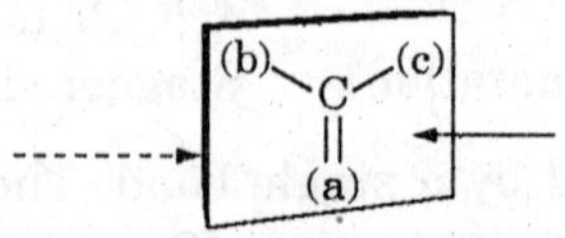

[B]

This face is *re* (above the plane of the paper). (a) → (b) → (c) clockwise movement.

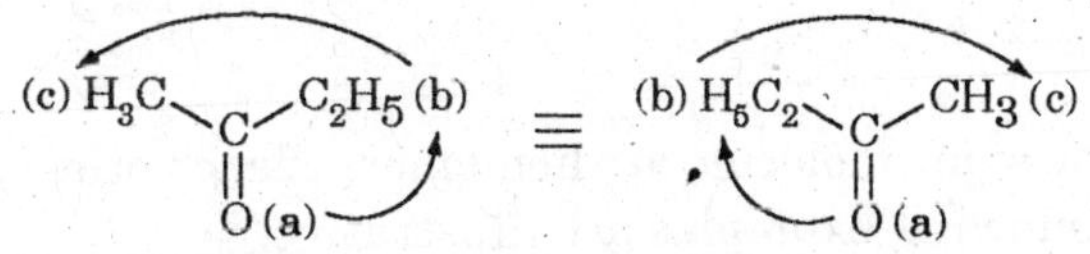

si-face (viewed from top) | *re*-face (viewed from under the paper)

specific, e.g., of [B]

[**Note**: When the front face is *re* the back face will be *si* and vice versa].

[iii] When the compound is of the type (a)(b)C_1=C_2 (b)(a) = [C], two faces are to be specified (one at C_1 and the other C_2) on each surface (i.e., front or back). Thus,
(By CIP, a → b → c)
(*E*-notation of [C]).
(From top (a) → (b) → (c_2) = *re*), (continuous arrow)
(From top (a) → (b) → (c_1) = *re*) (dotted arrow)
So the front face of [C] is *re–re*. Therefore, opposite face of [C] is *si-si*.

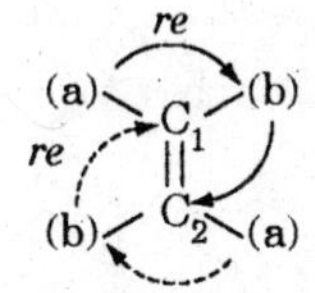

trans-compound
E-compound.
[C]

1. **Fumeric acid**: [D] is *trans* and E.

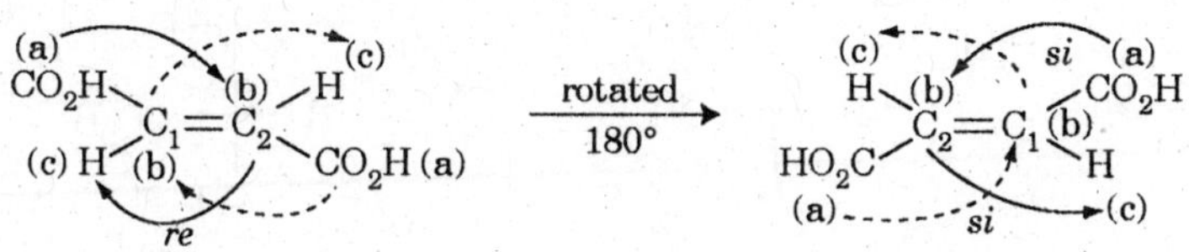

[D] | [D]

Front view (by CIP)
(a) → (b) → (c) at C_1 = *re*-face and
(a) → (b) → (c) at C_2 = *re*.
This is re-re face of [D].
(dotted arrow)

Backside view (by CIP)
a → b → c at C_1 = *si*-face and
(a) → (b) → (c) at C_2 = *si*.
This is *si-si*-face of [D].
(dotted arrow)

2. **Maleic acid**: *cis* and *Z* = [F]. Isomer of fumeric acid. This is stereoisomeric called diastereomer of fumeric acid and vice versa.

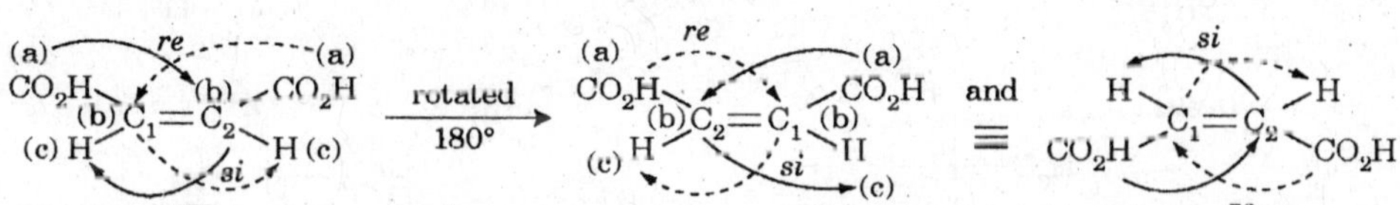

[F]

Front view (by CIP)
(a) → (b) → (c) = *re*-face at C_1.
(a) → (b) → (c) = *si*-face at C_2. (dotted area)
This is *re-si* face of [F].

Back side view (by CIP)
(a) → (b) → (c) = *si*-face. at C_1.
(a) → (b) → (c) = *re*-face at C_2. (dotted area)
This is *si-re* face of [F].

This is *si-re* face of [F].

45. Homomers

Different drawings of the same molecule are homomers of each other. They are superposable in every respect. The following examples are illustrative.

Examples

[i] 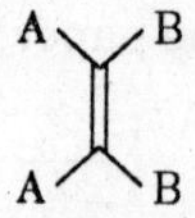 ≡ [ii] 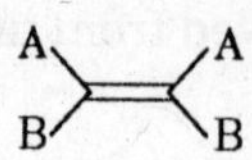 ≡ [iii] 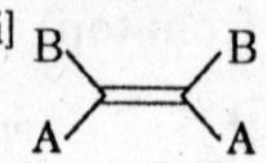 ≡ [iv] 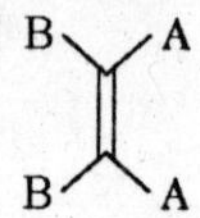

cis-$C_2A_2B_2$

(i) through (iv) are homomers of *cis*-$C_2A_2B_2$.

[v] 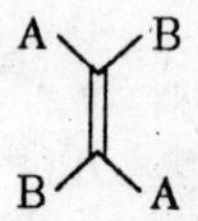 ≡ [vi] 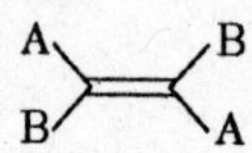 ≡ [vii] 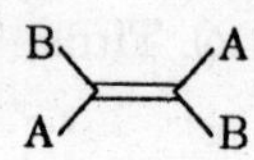 ≡ [viii] 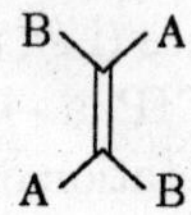

trans-$C_2A_2B_2$

(v) through (viii) are homomers of *trans*-$C_2A_2B_2$.

But *cis*-$C_2A_2B_2$ and *trans*-$C_2A_2B_2$ are diasteremoric to each other (i.e., they are geometric isomers or *cis-trans*-isomers).

[ix] 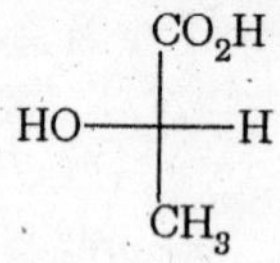

[x] 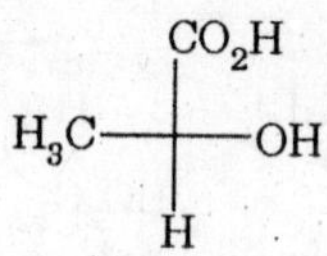

[xi] 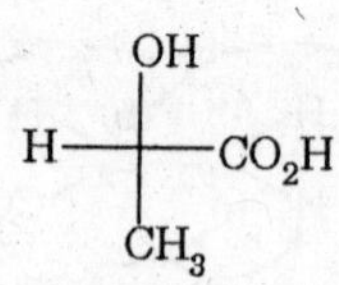

[xii] 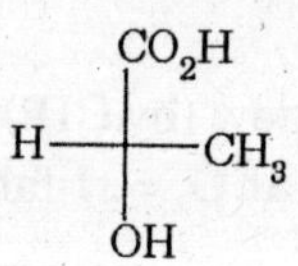

[xiii] 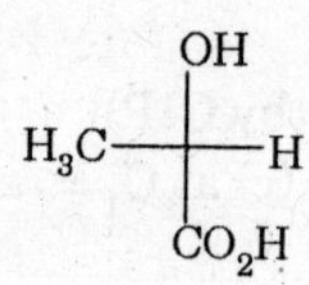

[ix] through [xiii] are homomers, (*S*-configuration).

[xiv] 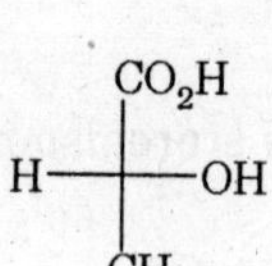

[xv] 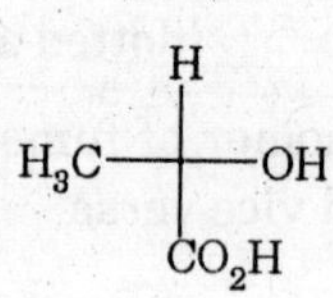

[xvi] 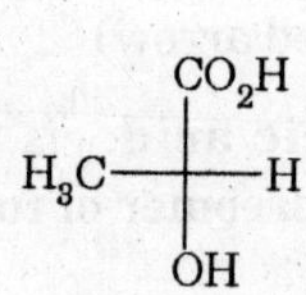

[xvii] 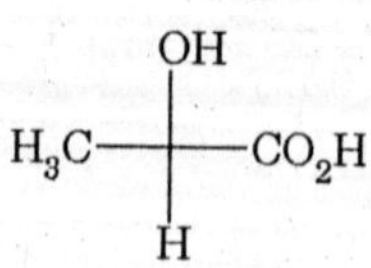

[xviii] 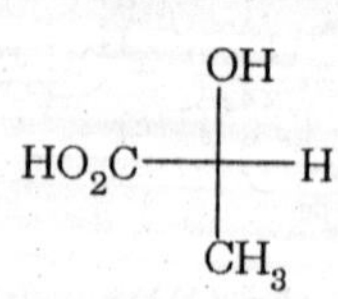

[xiv] through [xviii] are homomers, (*R*-configuration).

The drawings of (ix) through (xiii) series and (xiv) through (xviii) series are not homomers of each other, they are stereoisomers (optical isomers called enantiomers). They are mirror images of each other (non-superimposable).

46. Results of Transfer of Ligands

Transfer of three groups in a cyclic order will result into a homomer. Similarly, when two pairs are transferred among themselves the result is a homomer. Thus, (a) and (b) drawings are of same molecule *CABCD.

(a) 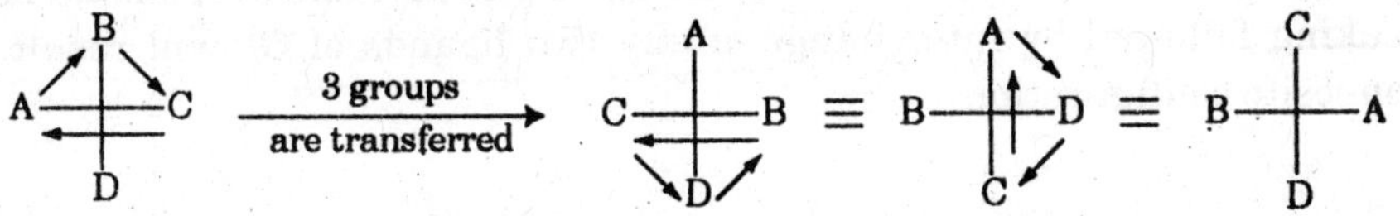 etc.

all are homomers

(b) 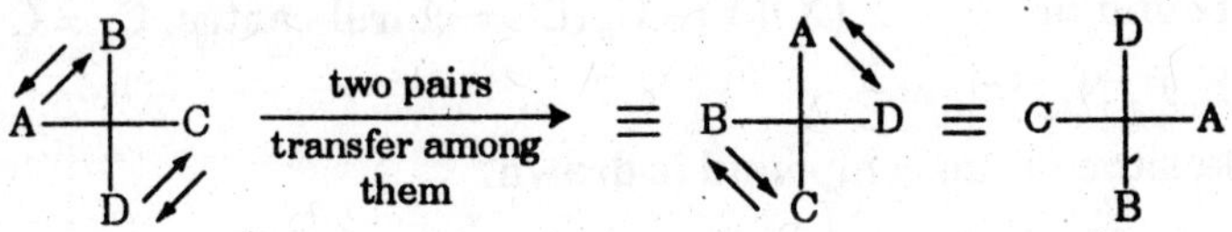 **all are homomers.**

But when only two groups i.e., one pair interchange their positions the result will be formation of an isomer. Similarly, if all four groups are changing the positions in cyclic order, then also an isomer is obtained. Thus,

(c) 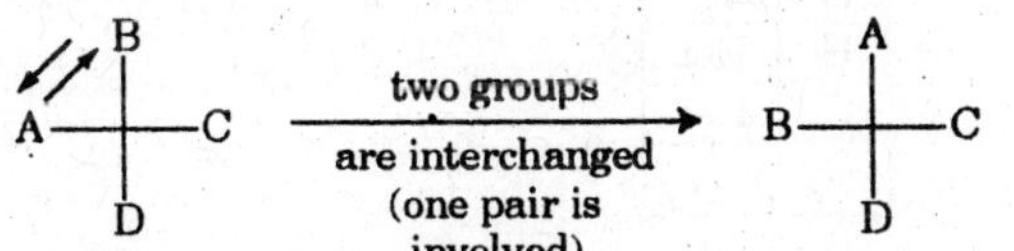

[I] [II]

These are isomeric (enantiomers). Not homomers. Not identical.

(d) 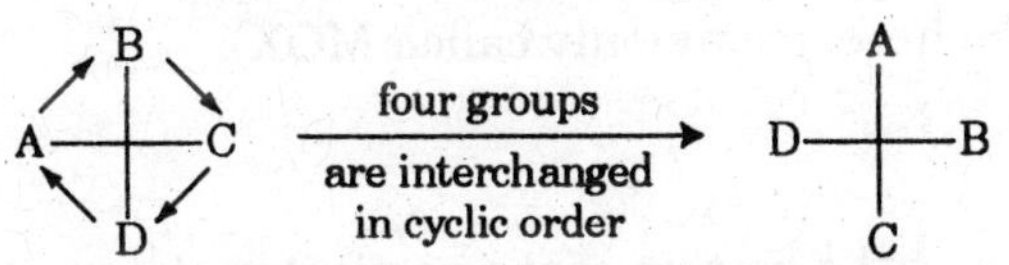

[III] [IV]

These are isomeric (enantiomers). Not homomers. Not identical.

[I] and [III] are idential. [I] and [II] are enantiomers. [III] and [IV] are enantiomers. [II] and [IV] are homomers.

Also,

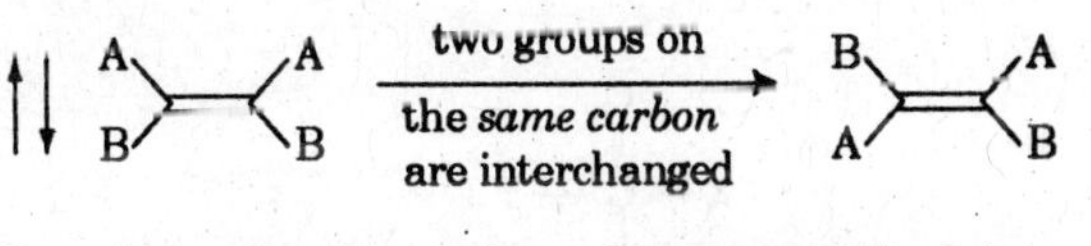

Result is the formation of a geometrical _isomer_ (diastereomer).

and 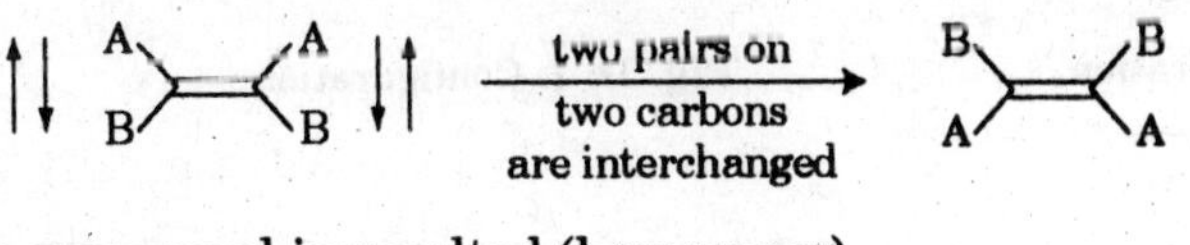

Same compound is resulted (homomers).

47. Assignment of the Configurational Notation to a Chiral Centre

The rigid arrangement of the atoms or substituents in space of a chiral carbon (i.e., a carbon which has four ligands different from each other, e.g., CABDE, this carbon is generally marked by asterisk, C*) is called the configuration of that carbon (C*). The notations of such configurations are D and L. But more sophisticated pair is *R* and *S*. It should be noted that bond breaking followed by interchange of any two ligands of C* will result in the formation of opposite configuration.

48. D/L Configurational Notations

Let the formula of the compound be $C_1 - C^*(X)(Y) - C_n$ (C^* = chiral centre, $C_1 \neq C_n$), and Het = heteroatom, e.g., O, S, P, N, Hal, etc.

[i] At first the vertical skeleton of the compound is drawn.

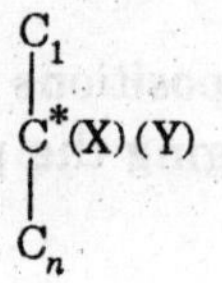

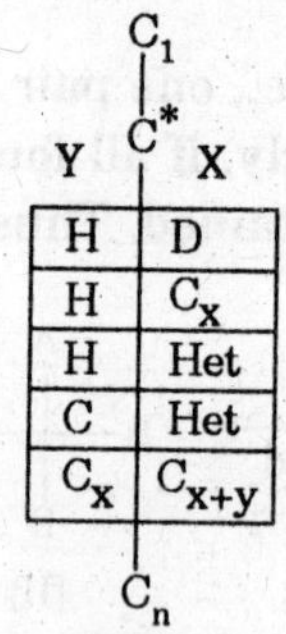

Fig. A

[a] In IUPAC name the C_1 will be on the top of the vertical skeleton, otherwise the most oxidised carbon atom will be here, (commonly called MOX).

[b] D in the box is deuterium.

[c] $C_{x+y} > C_x$ (groups).

[d] Less oxidised carbon atom will be at the bottom of the vertical skeleton, if C_6H_5 group (i.e., phenyl group) is present, it will come here, (commonly called LOX).

[e] D/L are relative and *R*/*S* are absolute configurations.

[ii] If the observed ligand under X be at right side of C* [Fig. A], then the notation will be D. On the other hand, if X is on the left-hand side of C* the notation will be L at this chiral carbon [Figs. B and D].

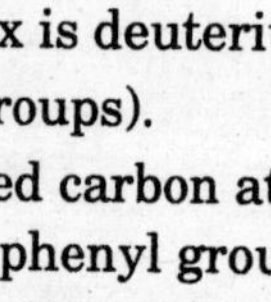

Fig. B: D-Configuration

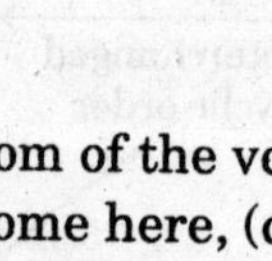

Fig. D: L-Configuration

Examples

[i] 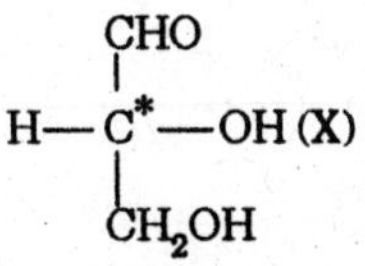

D-(+)-Glyceraldehyde

[ii] 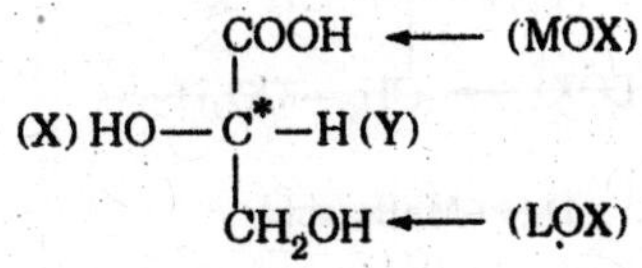

L-(+)-Glyceric acid

Oxidation state CHO > CH_2OH
X = OH = Het
Y = H. So D-Configuration

[a] D = notation (can be derived from structure).
[b] (+) means optically dextrorotatory (experimental, cannot be predicted from the structure).
[c] D and (+) are not related to each other.
[d] L but dextrorotatory.
[e] L and (+) are not related.

[iii] 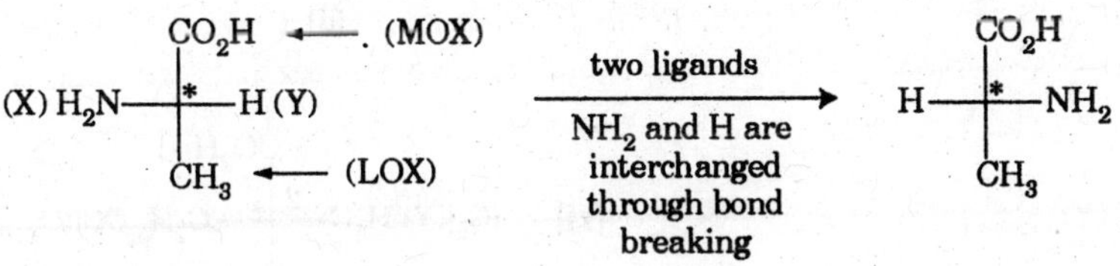

L-(+)-Alanine D-(–)-Alanine

[a] Configuration (D → L or L → D) changes; the optical rotation changes accordingly. [(+) → (–) and (–) → (+)].

[iv] 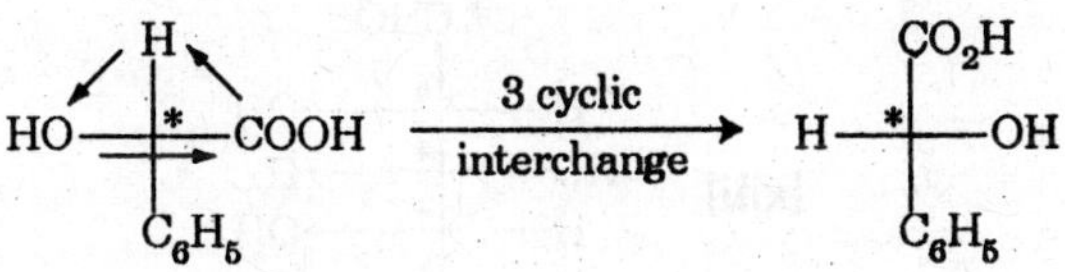

(–)-Mandelic acid [E] D-(–)-Mandelic acid [E]

[v] 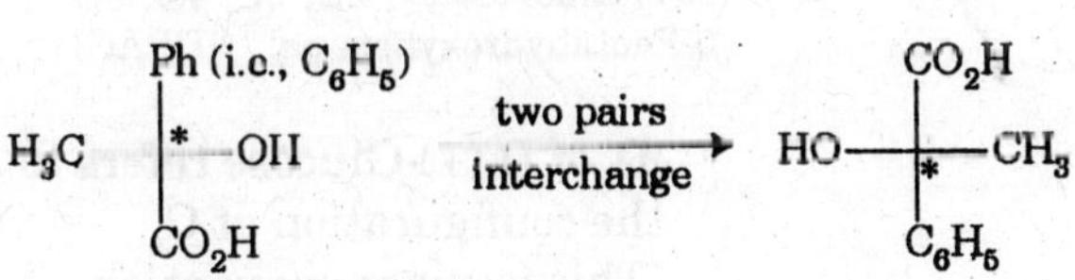

(+)-Atrolactic acid [F] L-(+)-Atrolactic acid [F]

[vi]

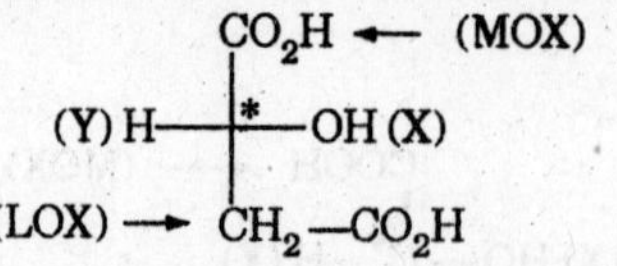

D-(+)-Malic acid
[G]

[vii]

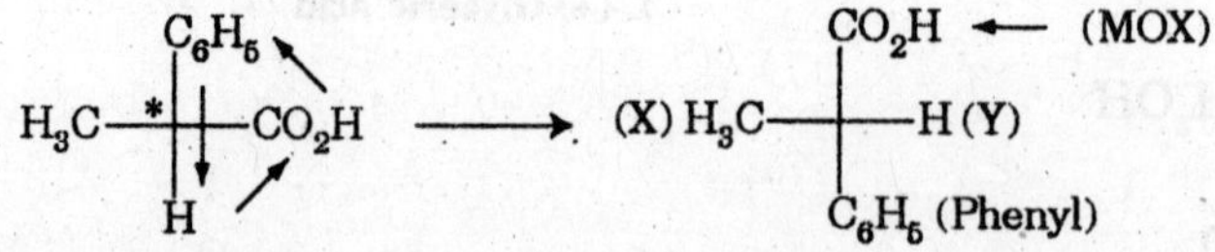

(+)-Hydratropic acid **[H]** — **L-(+)-Hydratropic acid** **[H]**

[viii]

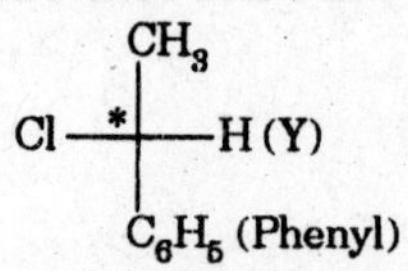

L-(+)-Chlorophenylethane
[I]

[ix]

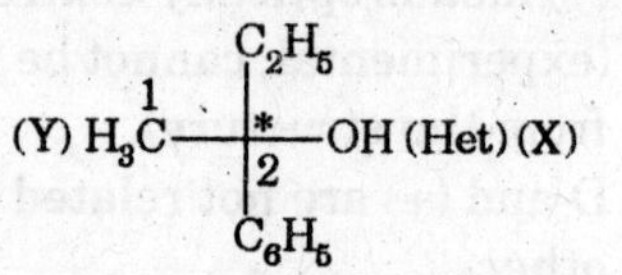

D-(−)-2-Phenylbutan-2-ol
[J]

[x]

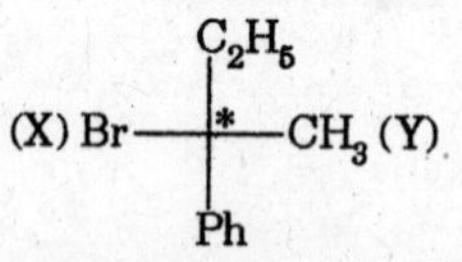

L-(+)-2-Bromo-2-phenylbutane
[K]

[xi]

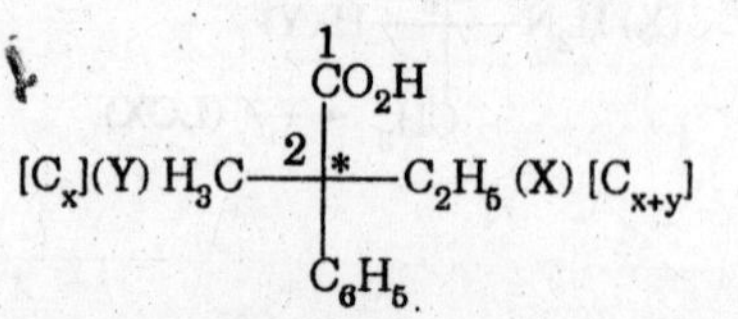

D-2-Methyl-2-phenylbutanoic acid
[M]

[xii]

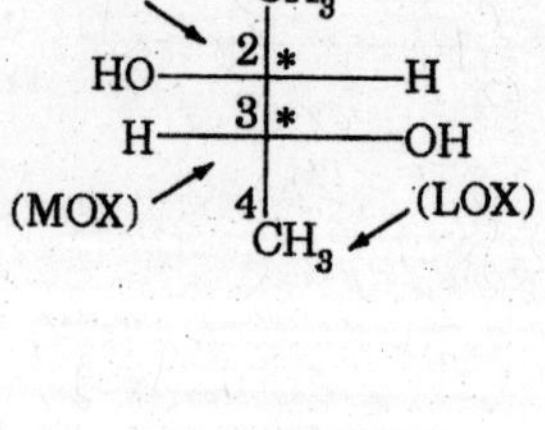

(−)-Butane-2D, 3D-diol
or
2D, 3D-(−)-Butanediol

[xiii]

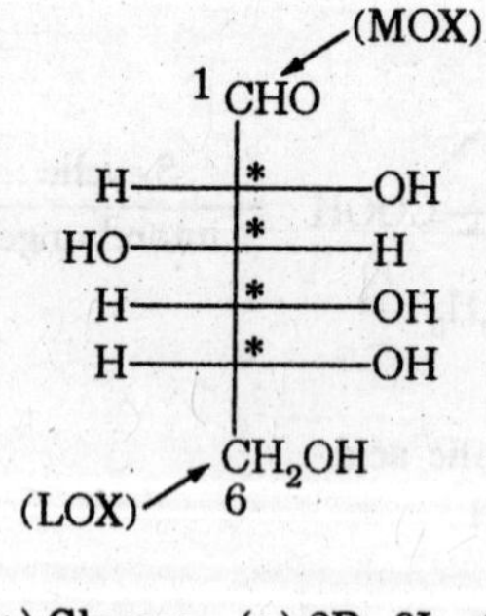

D-(+)-Glucose= (+) 2D, 3L, 4D, 5D, 6-Pentahydroxyhexanal. (IUPAC)

'D' of D-(+)-Glucose refers to the configuration at C_5. (This is sugar convention, the penultimate carbon determines the configuration.)

[xiv] 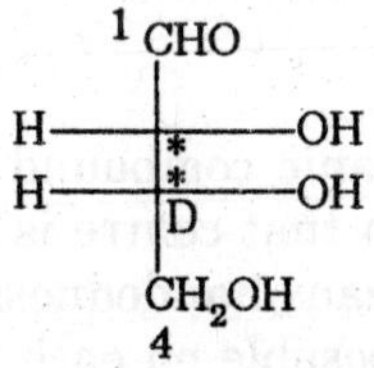

(–) 2D, 3D, 4-Trihydroxybutanal
[3D = Last but one carbon]
D-(–)-*Erythrose*

[xv] 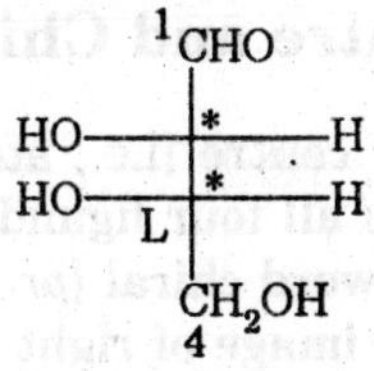

(+) 2L, 3L, 4-Trihydroxybutanal
L-(+)-*Erythrose*

Enantiomers

[xvi] 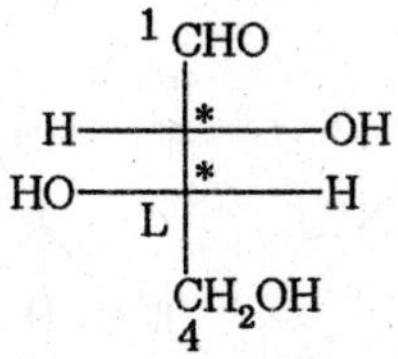

L-(+)-*Threose*
(+) 2D, 3L, 4-Trihydroxybutanal

[xvii] 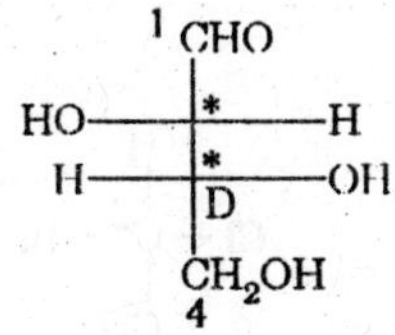

D-(–)-*Threose*
(–) 2L, 3D, 4-Trihydroxybutanal

Enantiomers .

49. Controversy on D/L-notations of (+) Tartaric Acid

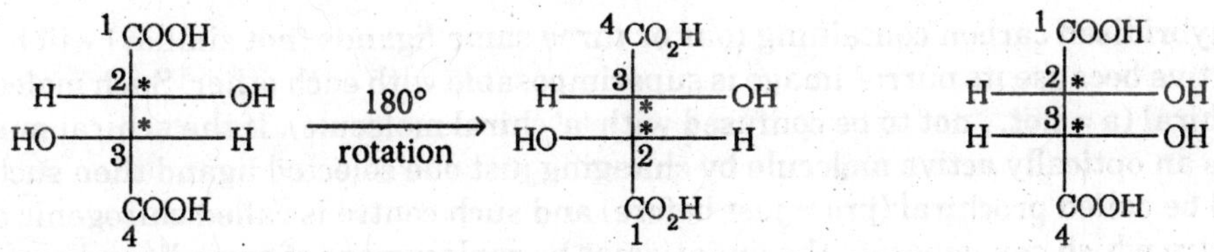

[A] (+)-Tartaric acid

[B]

2D, 3L-Tartaric acid (*meso*, σ, between C_2 and C_3)

C_2 = 2D notation (C_1 = MOX, C_3 = LOX, Het. right side, H = left side). Standard Procedure.

C_3 = 3L notation [if related to sugar, the Het. on lowest asymmetric carbon (C*) is on the left side, so 3L.] But,

C_3 = 3D notation (C4 = MOX but down, to apply standard procedure C_4 should be at top. This is done by rotation of Fig [A] through 180° to get Fig [B]. This converted figure shows the C_3* holding Het. (i.e., OH group) on the right, so 3D is the notation.)

Notes: [i] (+) Tartaric acid will be 2D, 3L by sugar convention but this becomes *meso* tartaric acid because 2D is the mirror image of 3L. But given example is optically active. Hence, if 2D is notation of C_2 (no doubt), the C_3 must be of 3D notation to make the given compound dextrorotatory [i.e., (+)].

[ii] Zirconium K_α – X-ray of $RbNaC_4H_4O_6$ supports the Fig [A] for (+) tartaric acid (Absolute configuration).

50. Chiral Centre and Chiral Molecule

When a particular centre [i.e., atom, generally carbon] in an organic compound is *sp*^{3} hybridised and has all four ligands different from each other, then that centre is called chiral centre. The word chiral [*pr.* Kairal; Greek Kheir = hand] means handedness. The left-hand is mirror image of right hand and they are not superimposable on each other. Generally the mirror image is imaginary picture of the real object but the mirror images of compounds posessing chiral centres are always real [i.e., both the object and its mirror image exist] and they are called *enantiomers* of each other which are nonsuperimposable, [C* = chiral centre]. This asymmetric carbon, C*, is not the 'Chiral Carbon' because carbon atom cannot be chiral but the centre is chiral.

Examples

[i] 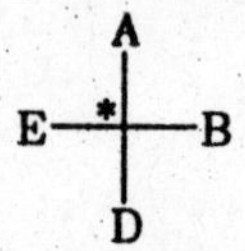

$A \neq B \neq D \neq E$
C* = Chiral Centre
(The molecule is asymmetric, optically active and chiral)

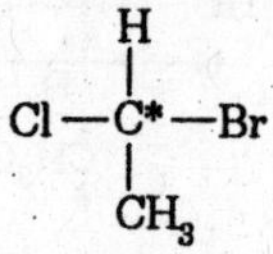

C* = chiral centre
The chiral molecule (Four ligands H_1, Br, CH_3 and Cl are different)

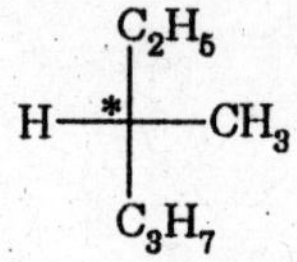

chiral molecule
(opt. active)
C* = Chiral Centre

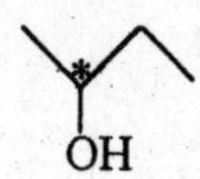

chiral molecule
(opt. active)

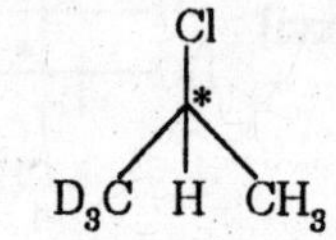

chiral molecule
($CH_3 \neq CD_3$)

51. Achiral, Prochiral and Chirogenic Compounds

The *sp*^{3} hybridised carbon containing four or three same ligands (not similar) will be optically inactive because its mirror image is superimposable with each other. Such molecule is called achiral (a = not, ; not to be confused with 'a' chiral molecule). If the achiral molecule generates an optically active molecule by changing just one selected ligand then such molecule will be called prochiral (pro = just before) and such centre is called chirogenic centre (i.e. a centre which can generate the enantiomer by replacement of same ligand one or the other).

Examples

[i] General

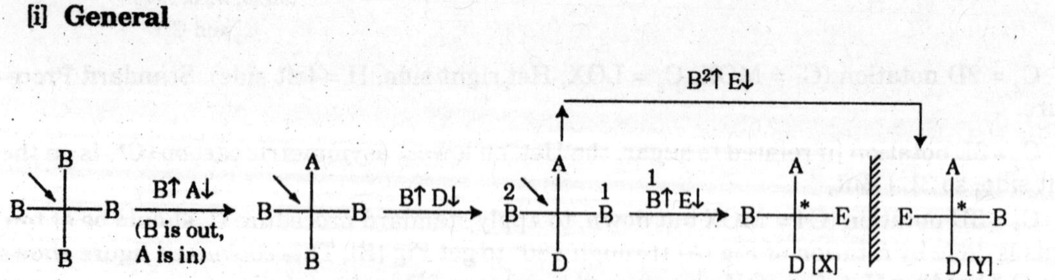

(Achiral molecule) (Four ligands are same)

(Achiral molecule) (Three ligands are same)

(Prochiral) centre marked by arrow.) ($B^1 = B^2$) *This prochiral centre is chirogenic.* (i.e., can generate chiral centre)

([X] and [Y] are enantiomers sc chiral molecules)

[ii] Specific

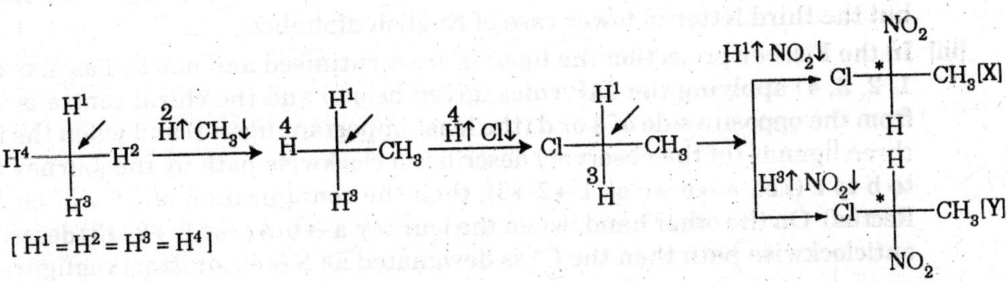

(Achiral) (Achiral) Prochiral (centre is chirogenic) [X] and [Y] are enantiomers and chiral molecules.

[It is to be noted that achiral and prochiral centres are not restricted in the sp^3 carbon. Such centres may be present at sp^2 carbons, but when the molecule becomes chiral the centre is always sp^3 hybridised.]

Examples

[i] 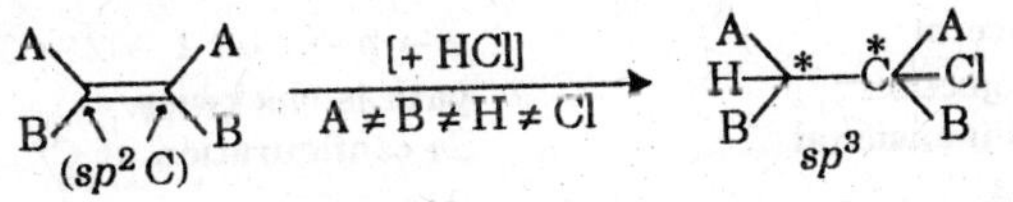

(prochiral centres marked by ↑) (C* = Chiral centres)

[ii] 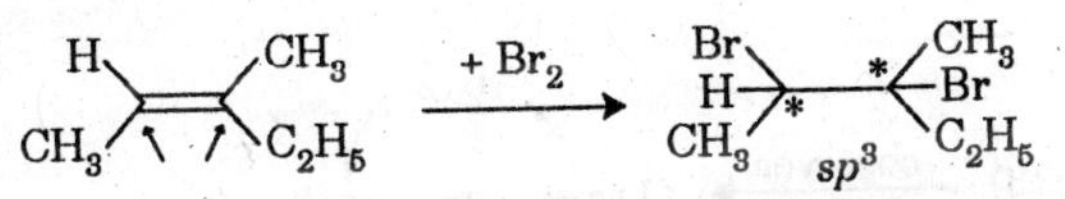

(prochiral centres) (chiral centres*)

[iii] 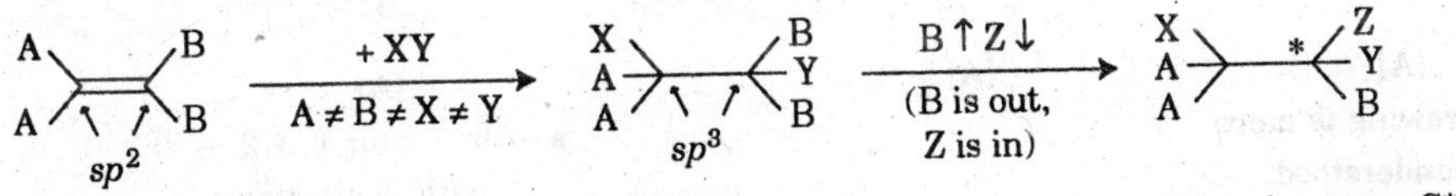

(prochiral centres) (prochiral) (chiral centre, C*)

52. *R*/*S*-Configurational Notations or Descriptors

The chiral centre of a molecule can be marked as *R* (from Rectus = right) and *S* (from Sinister = not right, i.e., left) which are more definite and unambiguous than D and L notations of the chiral centres. This *R*/*S* notation is derived by CIP sequence rules as shown below. The CIP is derived from the initial letters of the names of three chemists, Cahn, Ingold and Prelog who introduced these rules.

[a] Application of CIP Procedure

[i] First step is to identify the chiral centre(s) in the molecule.

[ii] Next step is the determination of priority of each one of the four ligands of each one of the chiral centres (C*). Then the priority order is written taking the most

'superior ligand as 'a' (or 1) and the least important ligands as d (or 4) like a>b>c>d or 1>2>3>4 (priority decreases in this order). The 'c' *in this order* is not carbon but the third letter in lower case of English alphabet.

[iii] In the Fischer projection the ligands are scrutinised and marked as a, b, c, d (or 1, 2, 3, 4) applying the CIP rules (given below) and the chiral centre is viewed from the opposite side of 4 or d (the least important ligand) and when the nearer three ligands (of the observer) describes a clockwise path by the journey from a to b to c (i.e., a→b→c or 1→2→3), then the configuration of C* will be *R* (i.e., Rectus). On the other hand, when the journey a→b→c (or 1→2→3) describes an anticlockwise path then the C* is designated as *S* (i.e., sinister) configuration.

- **General Examples**

[i] 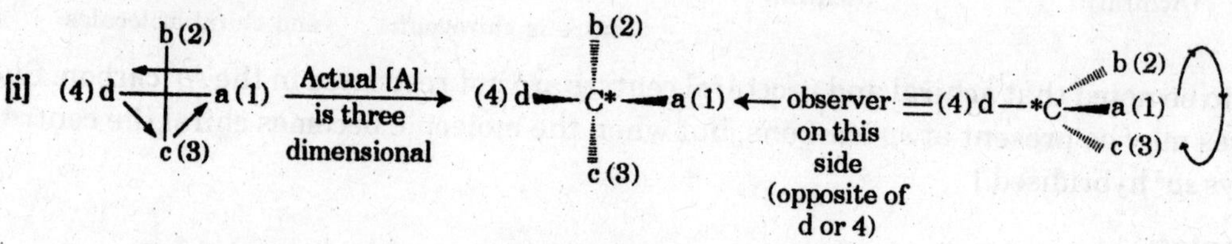

[A] Fischer Projection

Significance of Fisher Projection in three dimensional drawing [A].

a → b → c (or 1 → 2 → 3) path is clockwise. So configuration at C* is *R*[A].

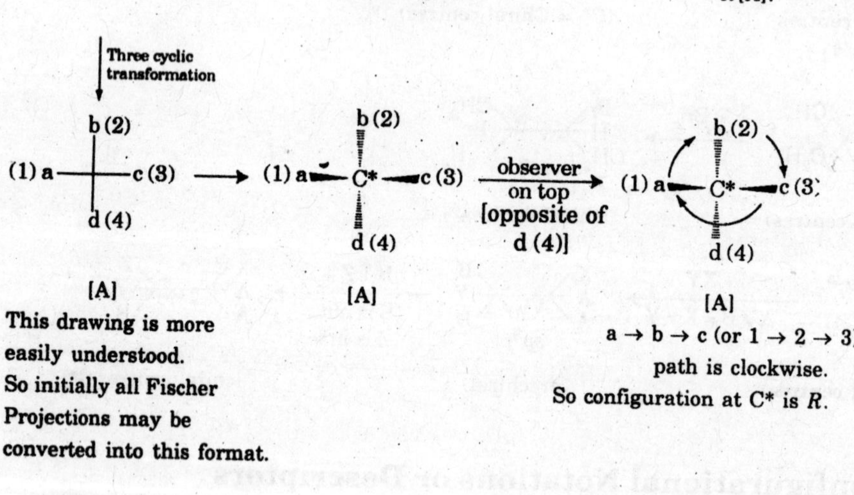

[A]

This drawing is more easily understood. So initially all Fischer Projections may be converted into this format.

[A]

[A]

a → b → c (or 1 → 2 → 3) path is clockwise. So configuration at C* is *R*.

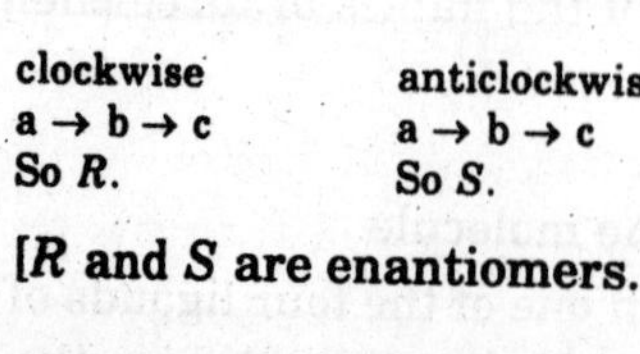

clockwise
a → b → c
So *R*.

anticlockwise
a → b → c
So *S*.

[*R* and *S* are enantiomers. They are not *Superimposable*.]

[b] Determination of Priority: Sequence Rule

[i] Rule (1): Greater atomic number [mnemonic GAN] of the atom *directly attached to the asymmetric carbon* [chiral centre, C*] gets greater priority, marked accordingly as a, b, c and d [or 1, 2, 3 and 4]. Thus,

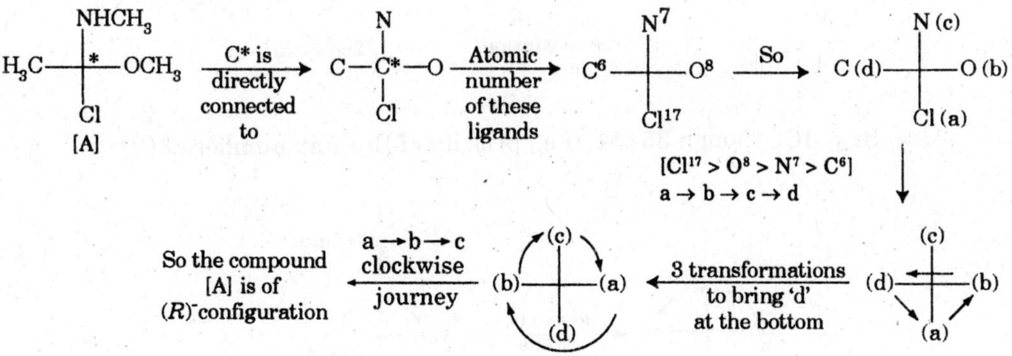

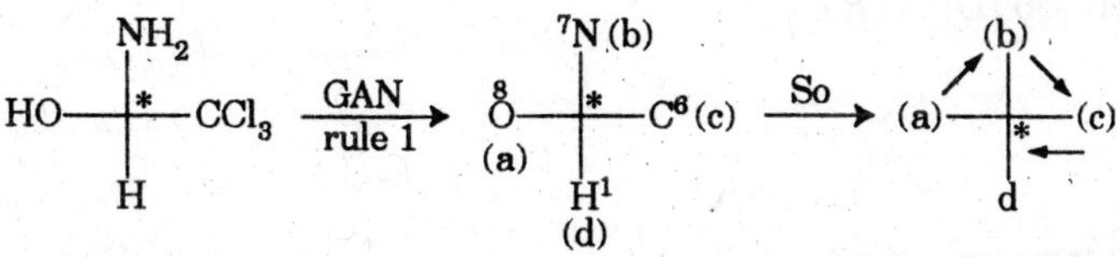

[B] (*R*)-configuration

So the compound [B] should be designated (*R*).

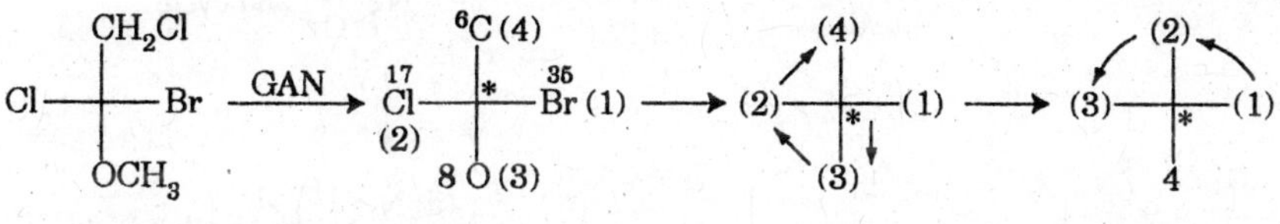

[C] (*S*)-configuration

The compound [C] is of (*S*)-configuration.

[ii] Rule (2): When the GAN (Rule 1) fails to answer the priority, then the next atom is considered as the candidate for greater priority [mnemonic BIGGAN]. If this second cycle (or step) cannot furnish the answer, then the next cycle (i.e., third cycle or step) is compared with each other. This will continue until the answer is obtained. The following points are to be carefully noted.

[a] This search for priority is to be stopped as soon as the answer is obtained and should never be extented to the last atom of the group in such case.

[b] It should be remembered that the atomic numbers are not to be added to determine the priority of a group. Thus one atom of greater atomic number has greater priority than two [even three] atoms of lower atomic number. Thus,

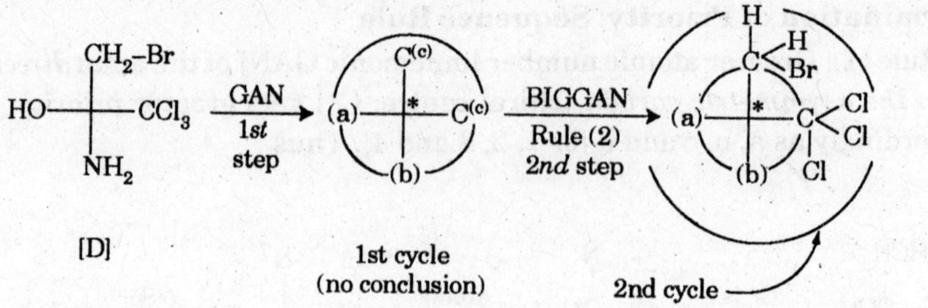

One $\overset{35}{\text{Br}} > 3\overset{17}{\text{Cl}}$ though 35<51. (i.e., priority of Br> any number of Cl).

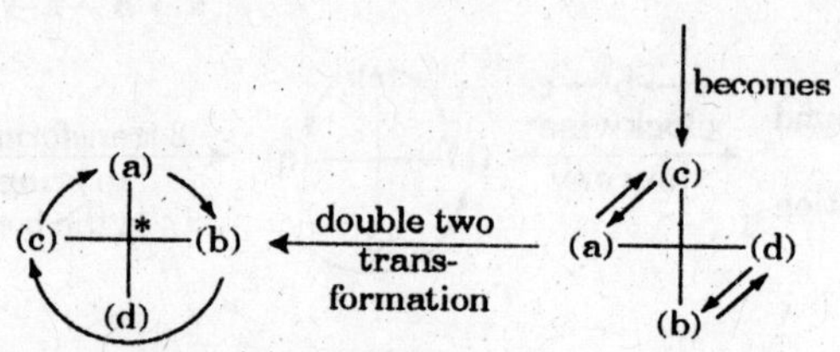

(*R*)-configuration. So [D] is (*R*).

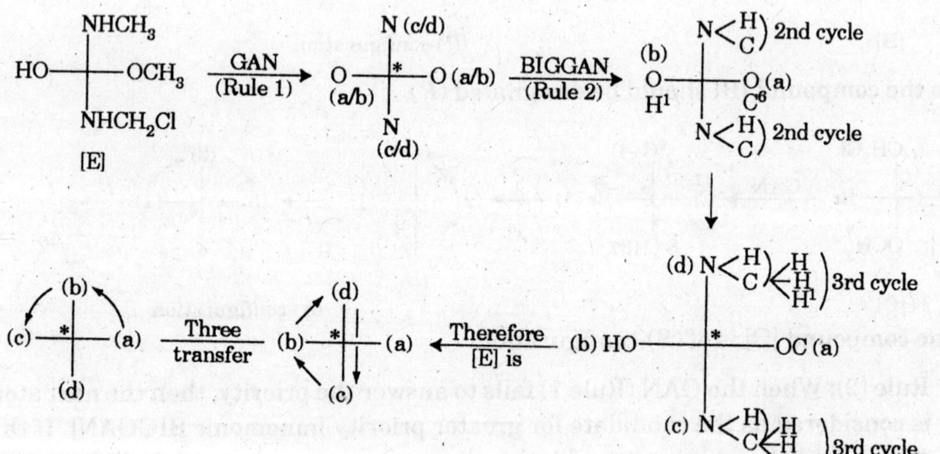

(*S*) configurations

[OCH_3(a), OH (b) $NHCH_2Cl$(c)$NHCH_3$(d)]

The priority determination is established in 3rd cycle. H, H H (in CH_3) Vs. H,H, Cl in (CH_2–Cl).

Revision of easy determination of *R*/*S* configuration.

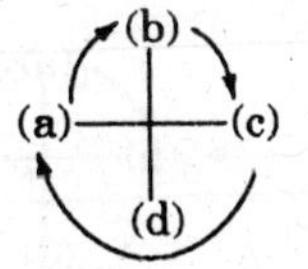

[R]

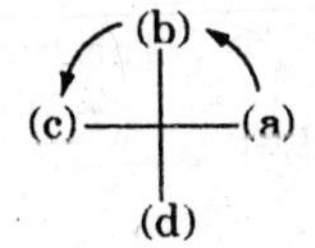

[S]

[i] It is to be noted that a little practice will enable the students to determine these configurational notations in one step.

[ii] When the *same element* occurs in different ligands, then the number of the element determines the priority, e.g. C,C,C>C,C,H>C,H,H>H,H,H i.e., $\overset{*}{C} — C(CH_3)_3 > CH(CH_3)_2 > CH_2CH_3 > CH_3$. But it has been already stated that a single element of higher atomic number in a particular cycle (of comparison) will get priority over two or more atoms of lower atomic number although the sum of the atomic number of the latter may be greater than the former one. Thus, $Cl^{17} > [O^8 ; O^8] > [N^7; N^7; N^7] > [C^6, C^6, C^6]$ and, $[N^7, H, H] > [C^6, C^6, C^6]$ etc.

Examples

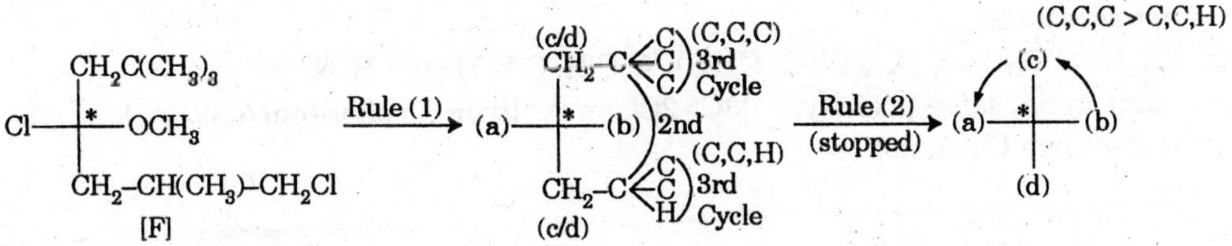

[F]

(S) configuration of [F]

Another way of priority depiction is as follows (F is taken as example).

				end		end
				C,C,C	3rd cycle →	C,C,H (d)
	end		end	H,H,C	← 2nd cycle	H,H,C
[F]	(a) Cl	← 1st cycle →	(b) O	C	;	C
	C* asymmetric; Carbon		C*	C*		C*

So, $Cl > O > C(CH_3)_3 > CH(CH_3)CH_2Cl$
(a) (b) (c) (d)

There is no need of fourth cycle scrutiny because the third cycle gives answer (i.e., which one is (c) and the CH_2—Cl group is ignored).

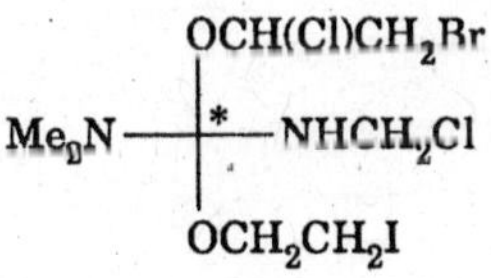

[G]

Priority cycle	end		end		end		end	
	H, C, Cl	3rd	H, H, C	3rd	end		end	
	C	2nd	C	2nd	H, C	2nd	C,C	2nd
	O	1st	O	1st	N	1st	N	1st
	C*		C*		C*		C*	
	(a/b)		(a/b)		(c/d)		(c/d)	

(H,C, Cl > H, H, C) in 3rd cycle (C,C > C, H) in 2nd cycle
(a) (b) (c) (d)

(The 4th cycle was unnecessary).
(The 3rd cycle was unnecessary).

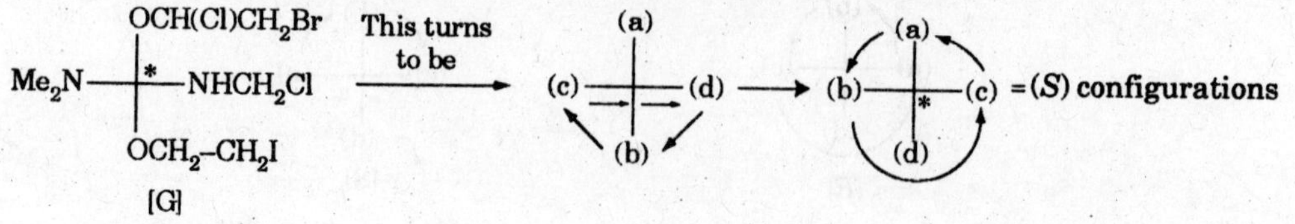

[G]

So compound [G] has *S*-notation.

[iii] Rule 3: When the chiral centre holds multiple bonded group(s), then the element forming multiple bond should be written as many times as the bond represents. One of them is true and remaining atoms are imaginary (false are called phantom atom). The true atoms and the phantom atoms (put in the parenthesis) have same priority in the same cycle but when shifted to the next cycle the phantom atoms are not considered (dead), only the atoms or groups connected to the true atom is considered for comparison.

Examples

Phantom atom (i.e., imaginary atoms satisfying the valency of the elements of first cycle) is given in parenthesis.

[a] C,C,C=C,C,(C)=C,(C),(C); C,(C),(C)>C,C,H; C,H,H>(C),H,H
[same cycle, same priority; at least one C is true] [3C>2C] [true C>phantom (C)]

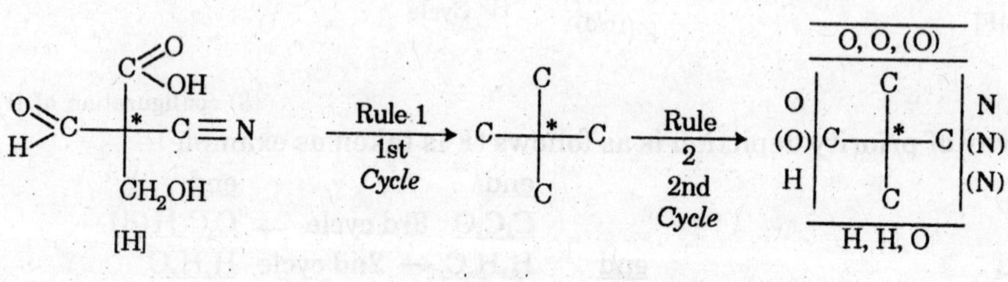

[H]

Priority

O(O)O>O(O)H>O,H,H N (N) (N)
(a) (b) (c) (d)

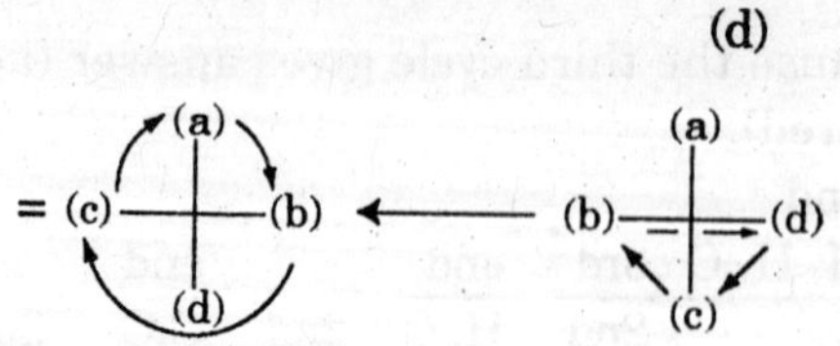

[R–] [H]

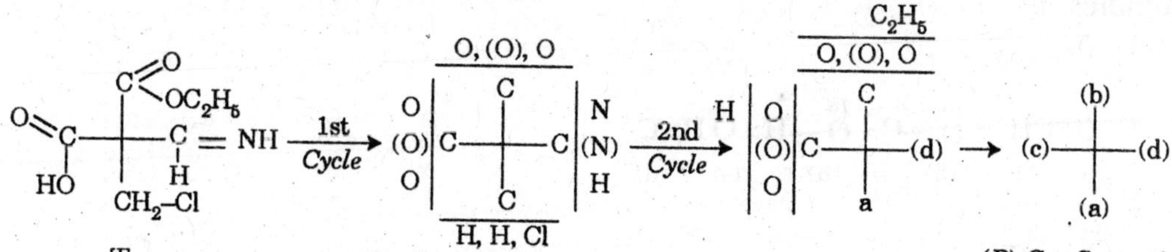

[I]

(*R*)-Configuration

H,H,Cl>N,(N) OC>OH
(a) (d) (b) (c)

[**Note:** The residual valency of C, N (and P), O (and S) in multiple bonds are three, two and one, respectively.]

[iv] Rule 4: When the chiral centre has only three ligands (atoms or groups), then the residual fourth valency (as a pair of electrons) is taken as (d) or 4, assuming that the fourth ligand is an imaginary element of atomic number zero.

Examples

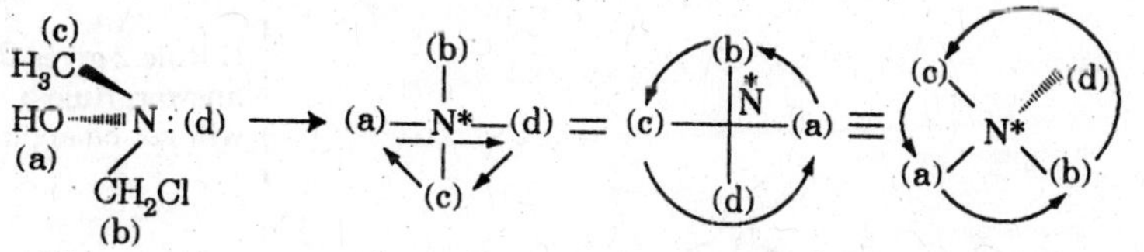

[J]

(*S*)-configuration (*S*)-configuration

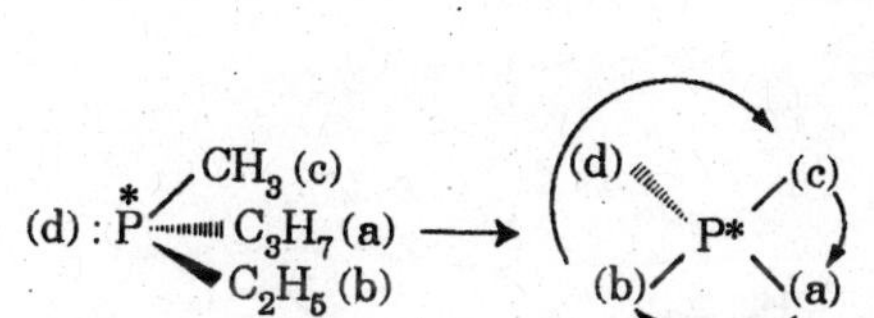

[K]

(*R*)-configuration; $C_3H_7 > C_2H_5 > CH_3 >$ (d)

[4] Quaternary ammonium salts with different four ligands should have the *R*/*S*-configuration.

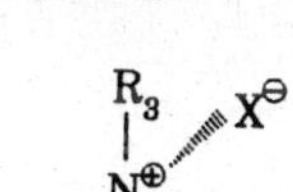

If priority $R_1 > R_2 > R_3 > X$, the salt will be of (*S*)-configuration.

[L]

[v] Rule 5: By the successive application of Rule 1 through 4 when the *R*/*S*-configuration cannot be determined, *only then* the atomic weight is considered. The greater atomic weight (GAW) gets greater priority.

Examples

Priority

$$O^{16}CH_3 = \overset{16}{O}\text{—}C > \overset{18}{O}\text{—}H > \overset{16}{O}H > C$$
$$\quad\;(a)\qquad\quad (b)\qquad (c)\qquad (d)$$

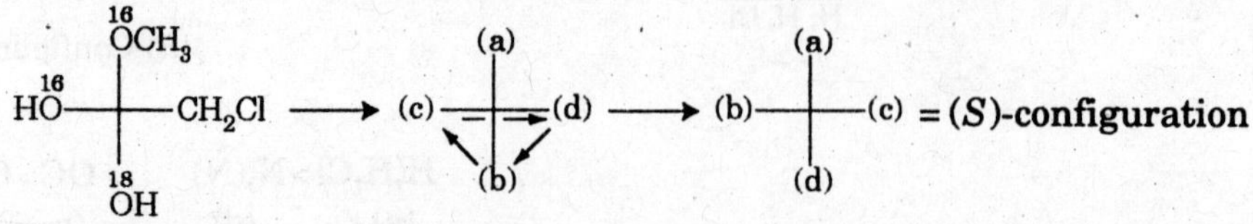

[M]

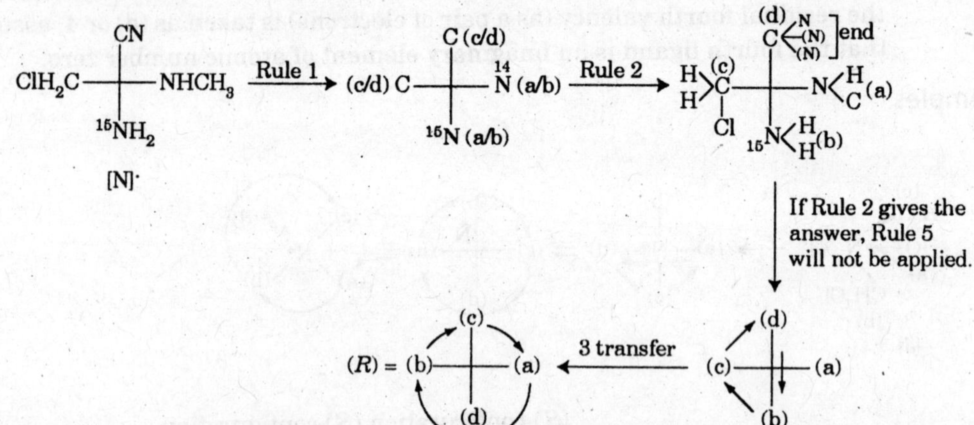

[N]

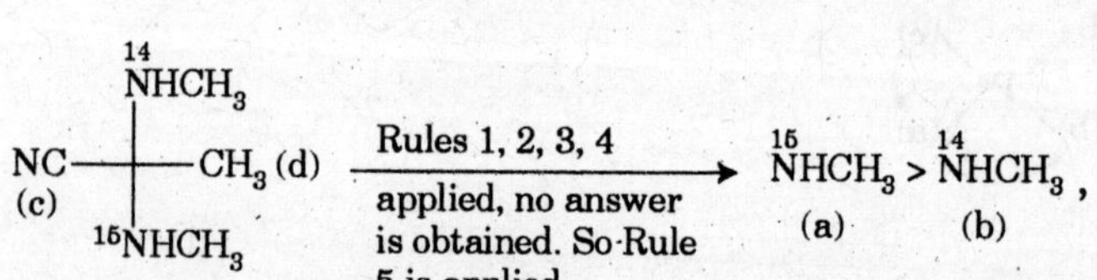

[O]

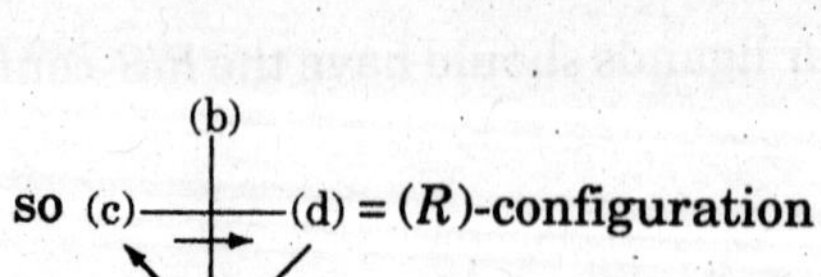

[Note: Since rules are applied stepwise, the— $CH_2–CH_2–\overset{18}{O}H > –CD_2–CH_2\overset{16}{O}H > CD_2–CH_2–CH_3 > CH_2–CD_2–CH_3$].

53. Different Projections of Organic Molecules

The alkynes are linear and alkenes are planar, they do not pose any problem to depict them on the plane surface. But the saturated carbon is three dimensional, hence one dimension has to be sacrificed while drawing their structures on the two dimensional plane surface.

This drawing can be done in four ways as shown below, taking $HO_2C-\overset{*}{C}H(OH)-CH_3$, as the example.

[i] *Fischer Projection*:

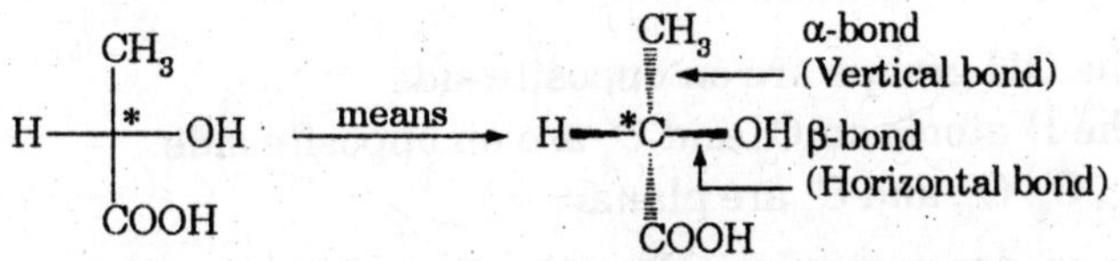

(*S*)-configuration [I]

[II] proposed as pseudo (i.e. ψ) Fischer projection in this book for ready reference to such drawings

[I] is correct Fischer projection.
[II] is wrong Fischer projection. (ψ–Fischer projection).
[I] and [II] are same compounds.

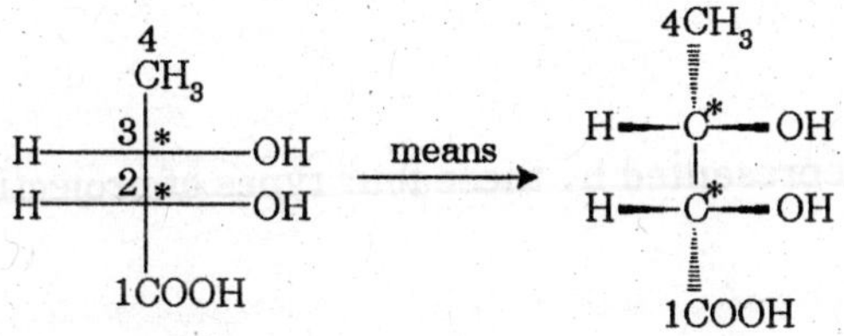

[III] [IV]

(This is akin to Flying-Wedge projection which shows α, β and on-plane bonds. So this is proposed as ψ-Fischer projection).

[III] is correct projection (Fischer).
[IV] is wrong projection (Fischer) (ψ-Fischer projection).
[III] and [IV] are same compounds.

[ii] *Newman projection*: The front asymmetric carbon eclipses the back carbon, the axis of vision passes through the asymmetric carbons. Thus III in Fischer projection looks like [V] in Newman projection.

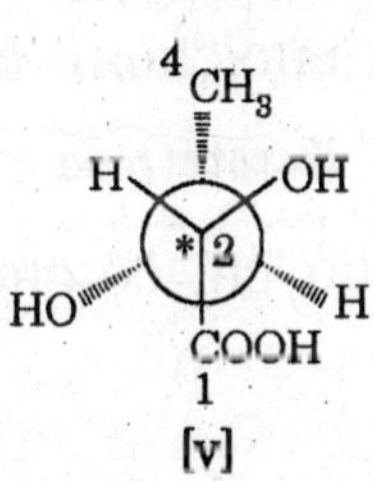

[v]

[a] C_2 eclipses C_3 (both asymmetric carbons).
[b] Axis of vision is vertical and passes through C_2 and C_3, the latter has been eclipsed by C_2.
[c] The angle shown is 120° between two bonds on the same carbon.
[d] Very suitable to demonstrate the conformational changes.

[iii] *Saw-horse projection*: This is the side way slanting drawing, the axis of vision is nearly 15° made with respect to the horizontal axis. Thus [III] in Fischer and [V] in Newman projections become [VI] in sawhorse projection.

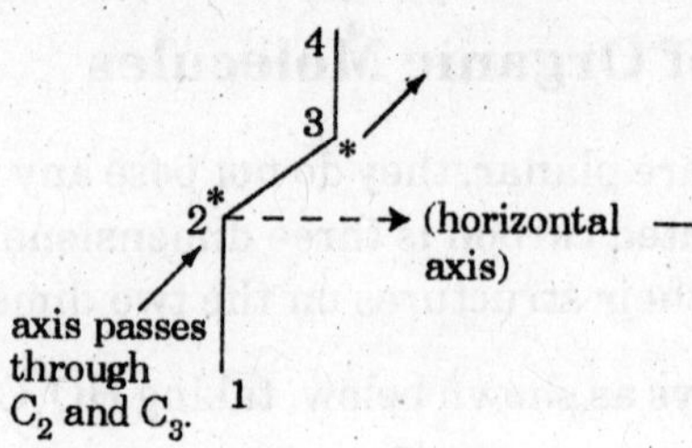

Skeleton of [VI]

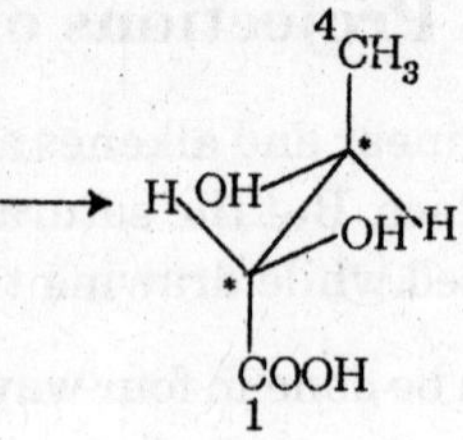

[VI]

[a] The OH groups are on opposite side.

[b] The H atoms on C_2 and C_3 are on opposite side.

[c] C_1, C_2, C_3, and C_4 are planar.

[iv] *Flying-wedge projection*: Here the C_2 and C_3 are drawn on the horizontal axis and substituents are shown as β- and α-bonds which are *anti* to each other. Thus [III], [V] and [VI] projections become [VII] in Flying-wedge projection.

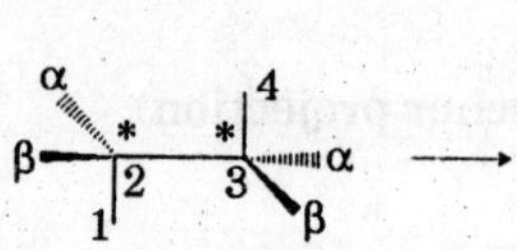

Skeleton of [VII]

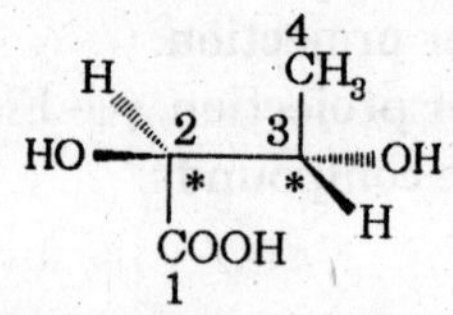

[VII]

[**Note**: Any organic molecule can be represented by these four types of projections].

54. Nomenclature of Sugars

These are saccharides, general formula $C_x(H_2O)y$; $x \geq y$ (exception, e.g., Rhamnose, $C_6H_{12}O_5$). The sugars are plant products and optically active compounds containing more than one OH group and a ketone or aldehyde function in the molecule.

[i] (a) The suffix of the sugar is –ose (*pr.* oze). (b) If CHO is present, it is aldose. (c) If keto group is present, it is ketose. (d) The number of carbons on the main chain (i.e., skeleton) is expressed as the multiplying prefix and connected with –ose.

Thus, $C_4H_8O_4$ [i.e., $C_4(H_2O)_4$] is Aldo‿tetra‿ose = Aldotetrose
(–CHO) (C_4) (Sugar) $CHO(CHOH)_2–CH_2OH$

and $C_4H_8O_4$ [i.e., $C_4(H_2O)_4$] is Keto‿tetra‿ose = Ketotetrose
($>C=O$) (C_4) (Sugar) $HO–CH_2–CO–CHOH–CH_2OH$

[ii] *Saccharides*: Sugars containing six carbons as the unit are called a monosaccharide and larger saccharides are derived from condensation of this unit by loss of water. Thus,

[a] $n \times$ monosaccharide $\rightarrow$ polysaccharide + $(n-1)H_2O$;
(n = whole number, > 1)

[b] $3 \times C_6H_{12}O_6 \rightarrow C_{18}H_{32}O_{16} + 2H_2O$; *oligosaccharide* contains C_{12}, C_{18} and C_{24}.
(n=3) Trisaccharide *polysaccharide*, $n > 4$

[iii] *D/L-notations in saccharide*: The penultimate carbon (last asymmetric carbon) determines the notation which has no relationship with optical rotation (i.e., +/– or *d/l*). The oxo group should be placed on top of the carbon skeleton and CH_2OH at the bottom. If the OH group in this frame is on the right side (of the observer), then the sugar is marked as D, otherwise it will be marked L (if –OH is on left), the OH under consideration is at the penultimate (i.e., last but one) carbon.

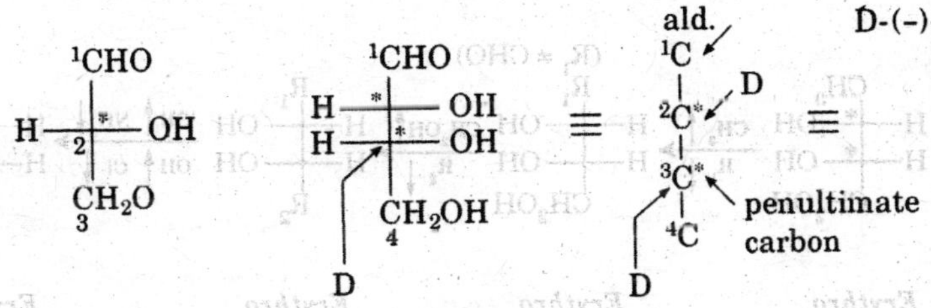

D-(–)-Erythrose
[D-(–)-Aldotetrosel]
≡(2D, 3D-Aldotetrose)

D-(+)-gylceraldehyde [D-(+)-Aldotetrose]

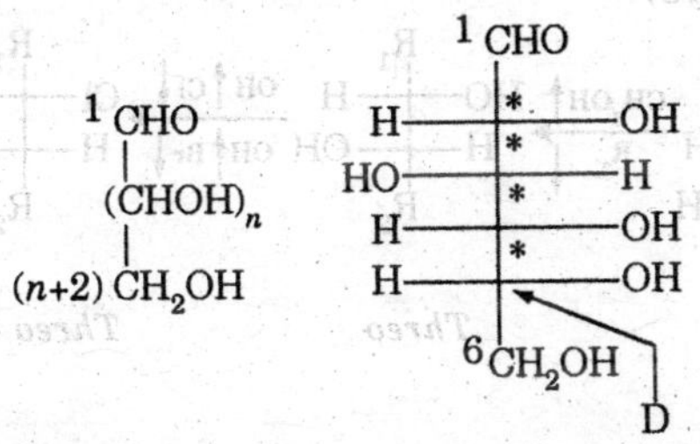

≡D-(+)-Glucose
[D-(+)-Aldohexose]; (n=4)
≡(2D, 3L, 4D, 5D, 6-Pentahydroxyhexanal)

[iv] The following examples are illustrative.

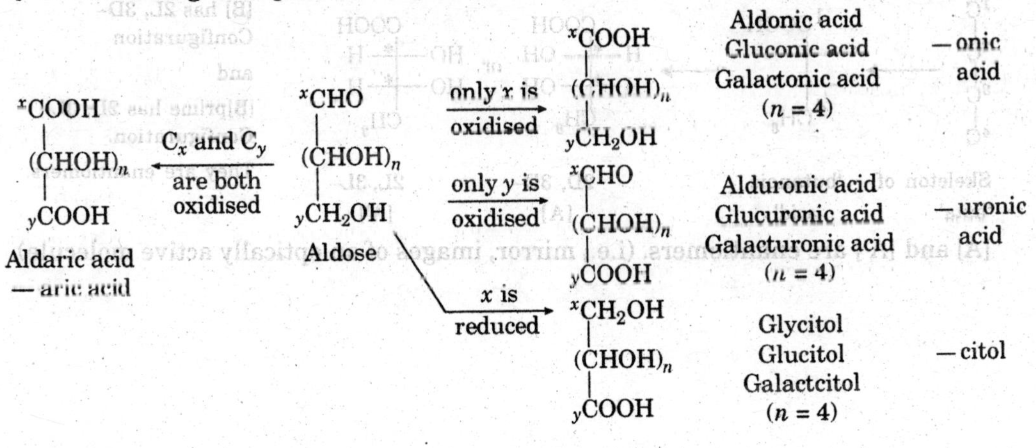

Aldaric acid $\xleftarrow[\text{both oxidised}]{C_x \text{ and } C_y \text{ are}}$ Aldose

Glucaric acid $\xleftarrow{\text{do.}}$ Glucose

Galactaric acid $\xleftarrow{\text{do.}}$ Galactose

55. [A] *Erythro/Threo* Configurations

Aldotetroses, $C_4H_8O_4$, are of two types, viz. erythrose and threose which are diastereomers of each other. When similar (or same) substituents are on the same side in Fischer Projection at C_2 and C_3, the compound will have *erythro*-notation and if these substituents (at C_2 and C_3) are on the opposite side in Fischer projection, it will be of *threo*-configuration. As soon as there is a deviation of the structure from sugar convention the suffix 'ose' is to be dropped. Thus,

[a] 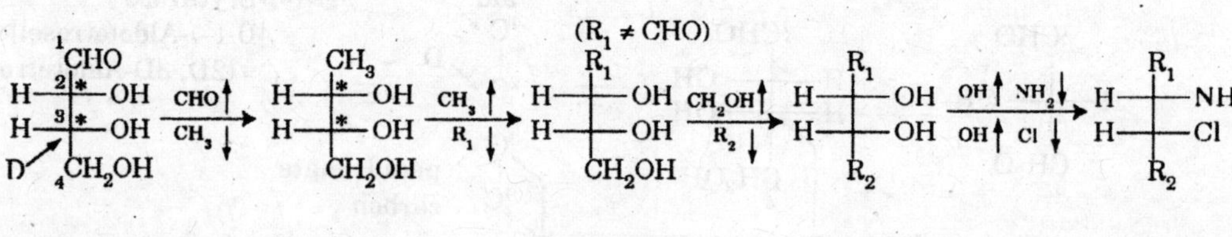

D-(–)-erythrose   *Erythro*   *Erythro*   *Erythro*   *Erythro*

[b] 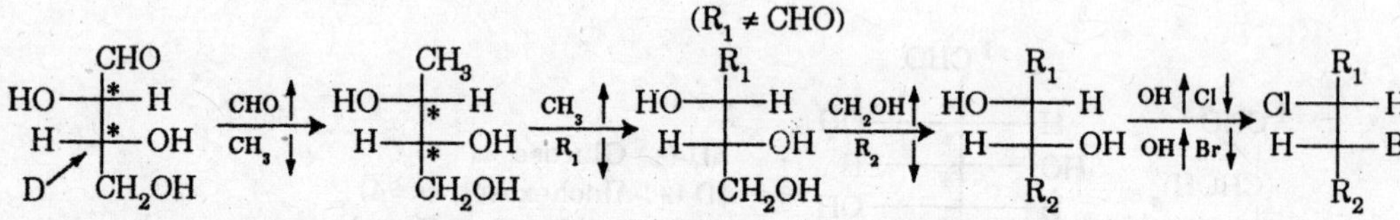

D-(–)-threose   *Threo*   *Threo*   *Threo*   *Threo*

[c] Other examples

1. *Erythro*-2,3–dihydroxybutanoic acid [A] (stepwise drawing).

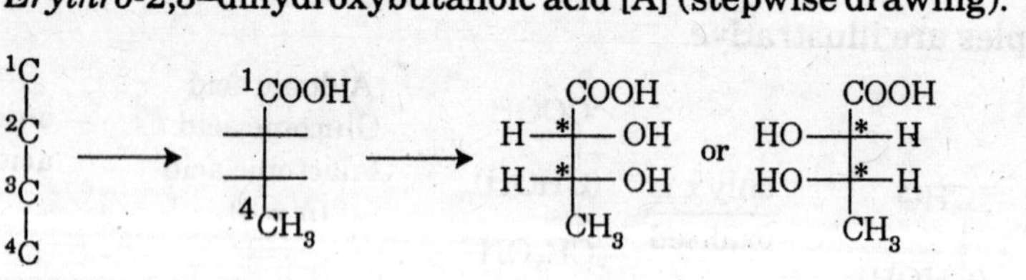

Skeleton of 'buta'   'butanoic acid'   2D, 3D– [A]   2L, 3L– [A′]

[B] has 2L, 3D–Configuration and [B]prime has 2D, 3L–Configuration. They are enantiomers.

[A] and [A′] are enantiomers. (i.e., mirror, images of an optically active molecule).

2. *Erythro*-2–ethyl–3–phenylbutanedioic acid [B]. (stepwise drawing)

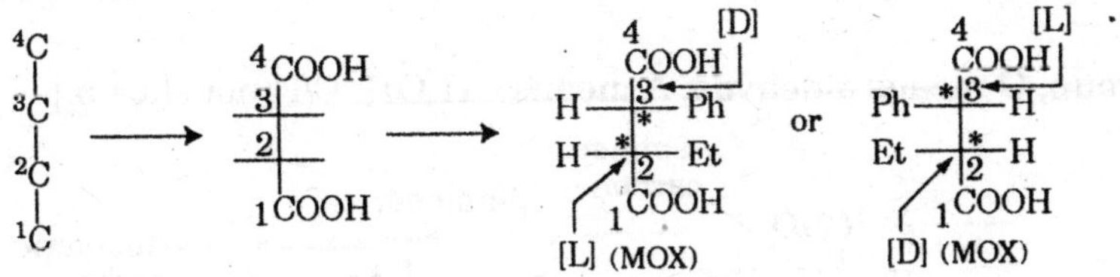

(butane Skeleton) | 'butanedioic acid' | [B] | [B]'

It is to be noted that *erythro* may lead to *meso*-compound and vice-versa.

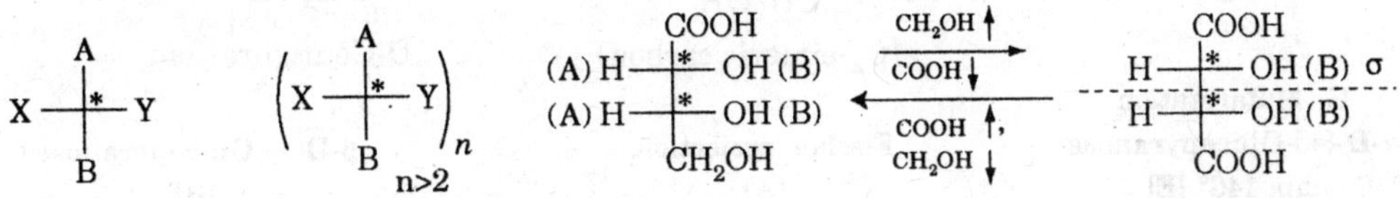

Not *erythro/threo* (as only one carbon is asymmetric) | Not *erythro/threo* (as more than two carbons are asymmetric) | *erythro* (Not erythrose, as this is not of sugar series) | *meso*-compound (Optically inactive since two halves, made by σ are mirror images

[B] Pyranose and Furanose Structures

[a] The sugars exist mainly as oxide rings in equilibrium with the open chain oxo-compounds. When the six membered oxide ring (engaging C_1, C_5 and O) and five membered oxide ring (engaging C_1, C_4 and O in aldose and C_2, C_5, O in ketose) occur, the sugars are termed accordingly with *pyranose* and *furanose* as suffix, respectively (e.g.F).

[C] Anomers

During the construction of pyranose rings the oxo groups turn into OH groups which may be on the same side of the oxygen of the oxide ring or on the opposite side (in Fischer projection), thus generating α-pyranose and β-pyranose structures, respectively. The α- and β-isomers of sugars are called *anomers*. In aldohexose the C_1 is the oxo group which generates the oxide ring of pyranose, so this C_1 is called *anomeric* carbon (e.g. F).

[D] Epimers

Isomerism at C_2 in aldoses generates epimers (e.g. F).

[E] Enantiomers and Diastereomers

The mirror image of an optically active compound and the object are *enantiomers*. The mirror image compound is also real like the object and they are non-superimposable. The *stereoisomers* which are not mirror images of each other are called *diastereomers*. When two or more *drawings* (i.e., representations) of the same compounds are *different*, they are *homomers*.

[F] Glycosides

When the anomeric OH group in sugar is converted into –OR (R=alkyl) group, then the resultant compound is the glycoside of that sugar.

[G] Mutarotation

Examples

[Let '—' means –OH group, O means aldehyde, A means CH_2OH, C/H not shown.]

[a]

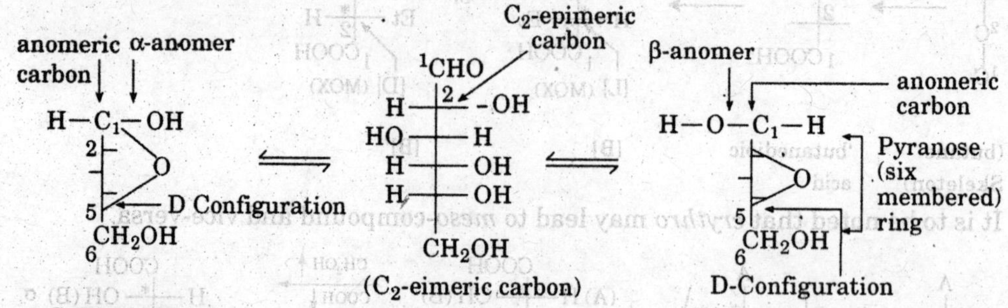

G. Mutarotation
α-D-(+)-Glucopyranose | **Fischer projection** | **β-D(+)-Glucopyranose**

mp. 146° [E] | **[A]** | **[B]**

$[\alpha]_D$ + 111° | | **mp 148–150°C**, **$[\alpha]_D$ + 19.2°**

In aqueous solution any pure anomer slowly equilibrates into the mixture of A, B and E, $[\alpha]_D$ + 52.20 (equilibrium mixture of A = trace; B = ca 64% and E = ca 36%). This phenomenon is known as *mutarotation*.

[H] The proposed abbreviation to make the Configuration more simple is as follows

In the Fischer projections of monosaccharides, the starting point is at D or L-representing carbon. If the OH group is on the right hand side (observer), it would be denoted by 0, (zero) and when OH is on the left hand side, it would be 1 (one). This will proceed upwards through the whole length of sugar. (This is just a temporary proposal). Thus,

D-(+)-Glucose =

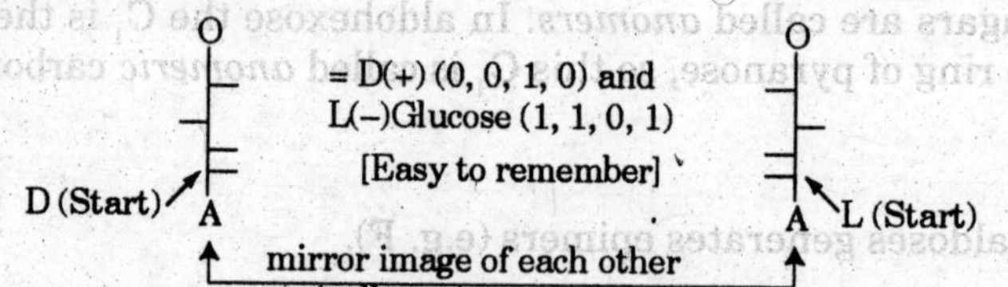

[c]

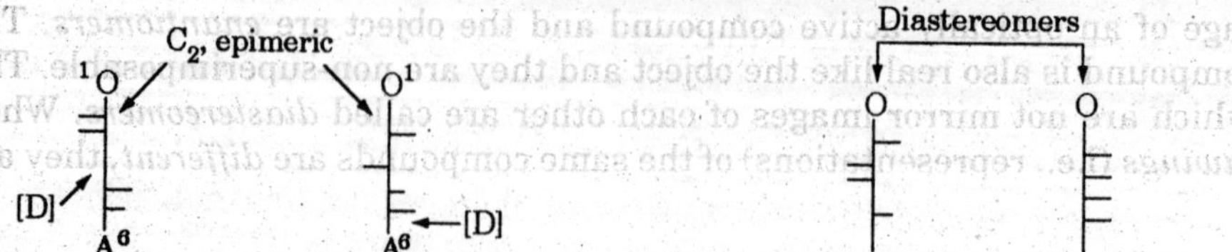

D-(+)-Mannose D-(+) (0, 0, 1, 1) | D-(+)-Glucose D (+) (0, 0, 1, 0) — epimers

D-(+)-Galactose D(+) (0, 1, 1, 0) | D-(+)-Allose D(+)(0, 0, 0, 0)

Note: *All* asymmetric centres when reversed will give *enantiomers*.
One or more (but *not all*) when reversed will give *diastereomers*.

56. Nomenclature of Aldoses, C_3 through C_6

The previous abbreviation will make the understanding of the constitution of the aldoses easy (the D series is taken as example and prefixes are given in parentheses).

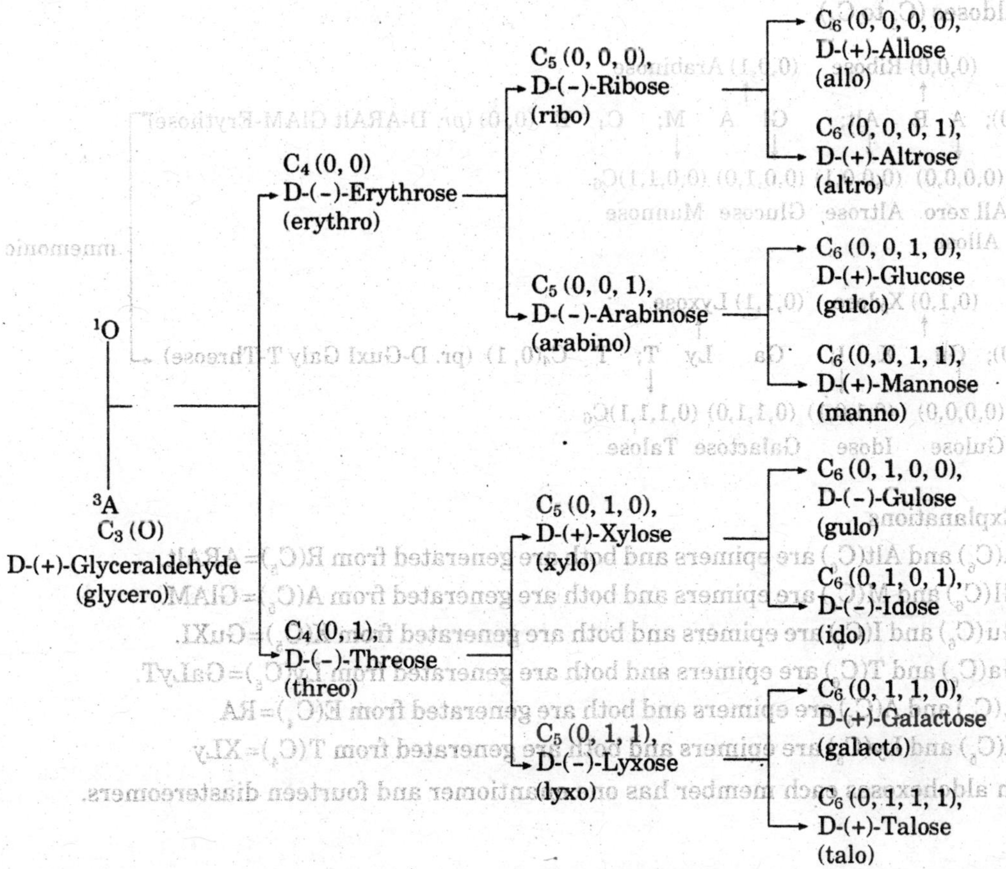

The following points are to be noted.

[a] Reversal of *all* the numbers (i.e., 0 to 1 and vice versa) will lead to the reversal of rotation sign generating its enantiomer. Example, C_5 (0, 1, 0), D-(+)-Xylose and C_5 (1, 0, 1) will be L-(−)-Xylose (they are enantiomers).

[b] Reversal of *one* number (or more but not all) will lead to the diastereomer with unpredictable optical rotation. The D/L configuration will change if the number of the penultimate carbon (i.e., the first number in the parenthesis) is changed (i.e., from 0 to 1 and vice versa).

[c] The optical rotations of the aldose cannot be predicted, but the following *mnemonic* may be helpful (for C_3 to C_6 only).

[D(+), D(–), D(–), D(+)]-except GuXI.[Gul(D–), Xyl D(+) and Ido D(–)]

↓ C_3 ↓ C_4 ↓ C_5 ↓ C_6 Gulose ↓ Idose ↓ Xylose

[d] The following will be also very useful *mnemonic* to recall to memory all names of the aldoses (C_3 to C_6).

(0,0,0) Ribose ↑ (0,0,1) Arabinose ↑

D (0); A R Alt; Gl A M; C_4 E (0, 0) (*pr.* D-ARAlt GlAM-Erythose)

↓ C_3 (0,0,0,0) (0,0,0,1) (0,0,1,0) (0,0,1,1)C_6

All zero Allose, Altrose, Glucose, Mannose

mnemonic

(0,1,0) Xylose ↑ (0,1,1) Lyxose ↑

D (0); Gu X I; Ga Ly T; T C_4(0, 1) (pr. D-GuxI Galy T-Threose)

↓ C_3 (0,0,0,0) (0,1,0,1) (0,1,1,0) (0,1,1,1)C_6

Gulose, Idose, Galactose, Talose

Explanations

A(C_6) and Alt(C_6) are epimers and both are generated from R(C_5)=ARAlt.

Gl(C_6) and M(C_6) are epimers and both are generated from A(C_5)=GlAM.

Gu(C_6) and I(C_6) are epimers and both are generated from X(C_5)=GuXI.

Ga(C_6) and T(C_6) are epimers and both are generated from Ly(C_5)=GaLyT.

R(C_5) and A(C_5) are epimers and both are generated from E(C_4)=RA

X(C_5) and Ly(C_5) are epimers and both are generated from T(C_4)=XLy

[e] In aldohexoses each member has one enantiomer and fourteen diastereomers.

57. Nomenclature of Carbohydrates larger than Aldohexose[18]

The carbohydrates containing more than four asymmetric carbons are named by more than one prefixes (vide 56) which will contain up to 4 asymmetric carbons indicating the configuration of the prefixes selected. These prefixes have the 'o' ending.

In Fischer projection the CHO group (=O) is taken on the top of the vertical sugar skeleton and CH_2OH (=A), ultimate carbon, is taken at the bottom of the skeleton. Then top four asymmetric carbons are scrutinised and appropriate prefix is taken. The D/L notation is determined on this part of the first prefix.

The remaining carbons determines the next prefix in the same manner with the proper D/L notation of the second prefix.

The total number of carbon atoms determines the multiplying prefix and connected directly with the suffix 'ose'.

The order of prefixes starts from bottom and proceeds upward (C/H not shown, '—' is OH group).

Examples (C/H not shown)

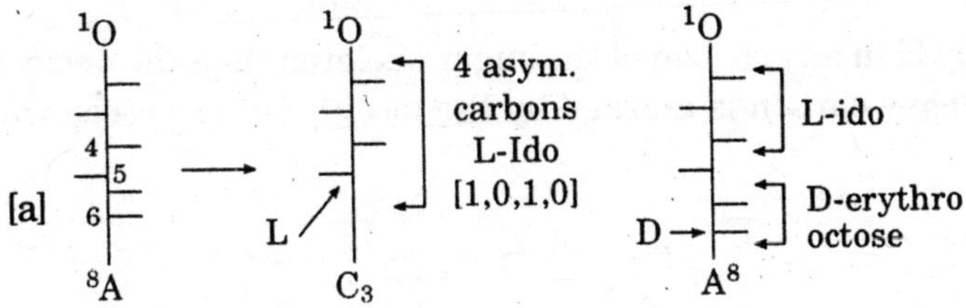

given compound C_8 = octa

Nom: [a] D-Erythro-L-idooctose—(Relative nomenclature)

[b] 2R, 3S, 4R, 5S, 6R, 7R, 8-Heptahydroxyoctanal (IUPAC). [A]

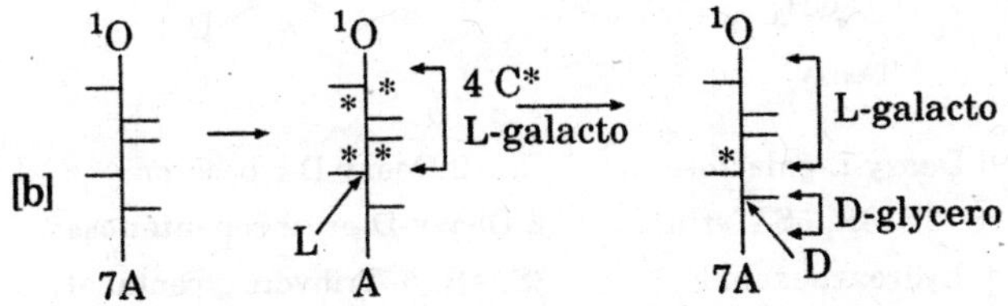

heptose (A=CH_2OH)

Nom: [a] D-Glycero-L-galactoheptose (Relative name)

[b] 2S, 3R, 4R, 5S, 6R, 7—Hexahydroxyheptanal.

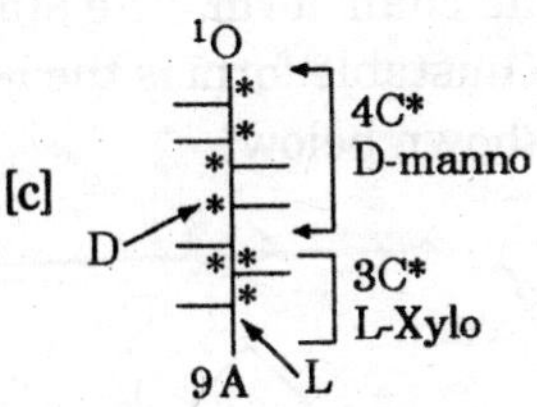

nonose (Nana+ose)

Nom: [a] L-Xylo-D-mannononose (Relative)

[b] 2S, 3S, 4R, 5R, 6S, 7R, 8S, 9—Octahydroxynonanal.

58. Nomenclature of Ketoses

These sugars have the suffix 'ulose' and the oxo groups have locants if they are not at C_2 (C_2 oxo does not require locant). The other procedure is just like that of aldose.

Examples

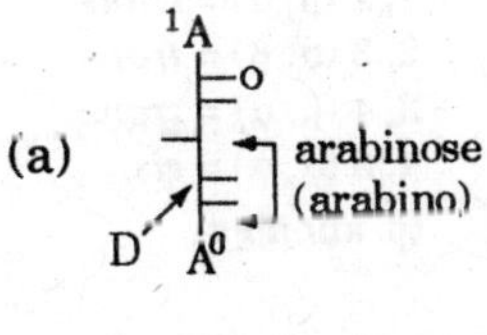

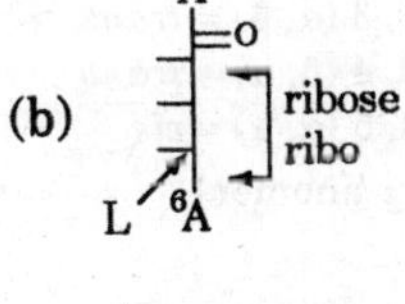

D-(–)-Fructose

D-(–)-Arabinohexulose

(1, 3S, 4R, 5R, 6-Pentahydroxyhexan-2-one)

L-Ribohexulose (six C chain = hexa + suffix, ulose)

[L-(–)-Psicose]

59. Deoxy Sugars

When the OH group is replaced by H in any carbon of the sugar skeleton then the deoxy (or desoxy) sugar is obtained. The deoxy carbon is marked by the locant. Other procedure of naming remains same.

Examples

(a)
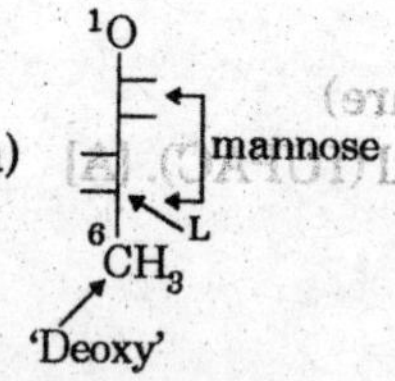

6-Deoxy-L-mannose
2R, 3R, 4S, 5S–Tetra-hydroxyhexanal

(b)
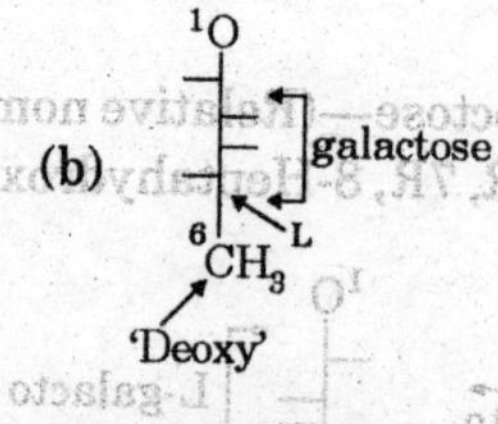

6-Deoxy-L-galactose
2S, 3R, 4R, 5S–Tetra-hydroxyhexanal

(c)
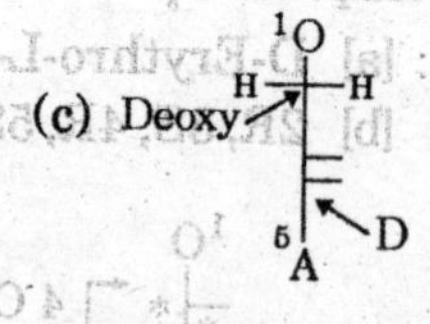

2-Deoxy-D-ribose *or*
2-Deoxy-D-erythropentanose
3S, 4R, 5–Trihydroxypentanal

60. Chair Form of α- and β- Anomers of D-(+)-Glucopyranose

The glucopyranose is α derivative of pyranose and it exists in the chair form. The stable stereochemical structure of cyclohexane is the chair form and the unstable form is the boat form. Likewise the glucopyranose is stable in the chair form as shown below:

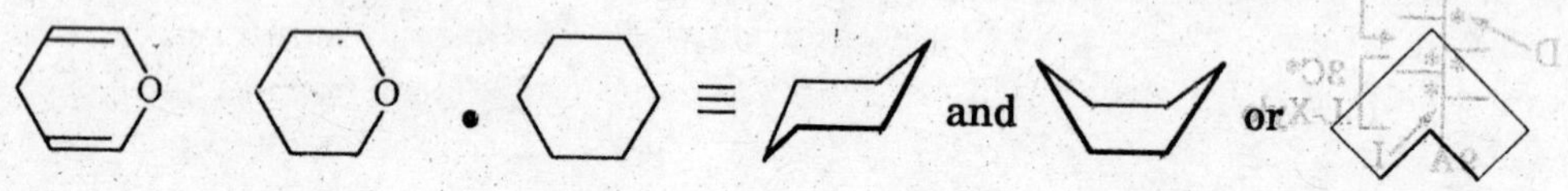

Pyran | Tetrahydro-pyran | Cyclohexane | Chair form | Boat form | hut form

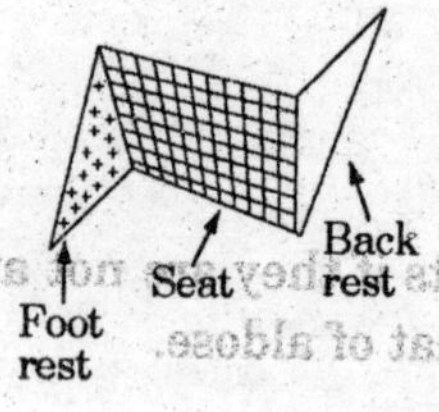

Chair illustrated

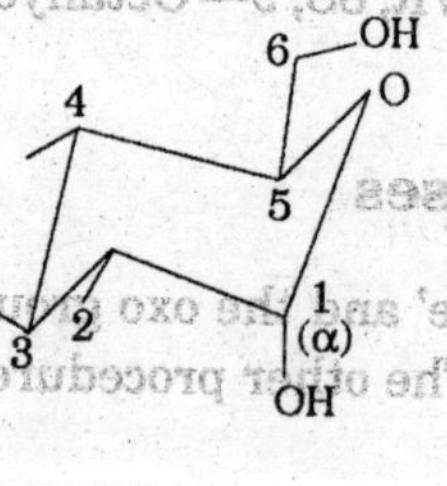

'—' is OH gr.
1,2 (α, α) = *cis*
2, 3 (α, β) = *trans*
3, 4 (β, α) = *trans*
4, 5 (α, α) = *cis*
(α-anomer)

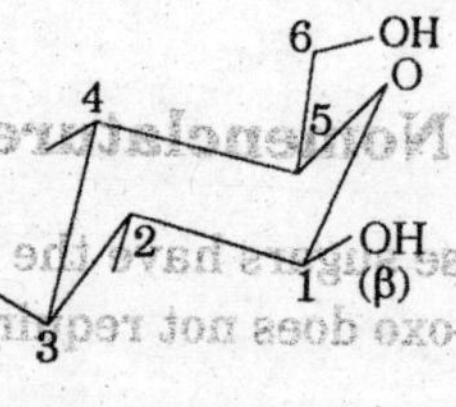

'—' is OH gr.
1, 2 (β, α) = *trans*
2, 3 (α, β) = *trans*
3, 4 (β, α) = *trans*
4, 5 (α, α) = *cis*
(β-anomer)

61. Locants of the Functional Groups

When only one functional group is present in the molecule, then generally it becomes the minor suffix as the termination of the name (one of the exceptions is NO_2 group which always becomes prefix). The root name is derived from the longest carbon chain containing the functional group. The major suffix of the root name is any one of 'ane', 'ene', 'yne' or 'enyne'. Whenever the functional group has to be used as prefix of the root name, then the locant (i.e., location of attachment with the carbons of the main chain) must be given. Otherwise the minor suffix (i.e., the functional group) is considered to have been attached (if the locant of it has not been specified) at C_1. But if there is a possibility of generation of isomers, then always the locants of functional group is to be mentioned. The rule of lowest sum and the rule of the lowest number is always followed taking the advantage of the mnemonic (LAAM LoRoo, Su, Pre, Co, Da, Vo, No) which has been discussed before (vide Sec. 8).

Examples

Ethanol (no locant), dimethylpropanol (no locant), propan-1-ol and propan-2-ol (isomeric), 1-nitropropane and 2-nitropropane (isomeric), cyclohexanol, 2,2-dimethylbutane and 2,3-dimethylbutane (isomeric) but methylbutane, etc.

62. Nomenclature of Compounds having two or more Functional Groups

When two (or more) functional groups are present in the compound, then only one of them has to be taken as the suffix and the remaining substituents are taken as the prefix(es). To choose the correct suffix (the minor one) poses problem. One way to avoid the problem is to consider all of the substituents as prefixes with the only major suffix from 'ane, ene, yne or enyne'. This procedure may be a way out (see example) but the naming of the compound may not be the best. However, 'all prefix' naming procedure will always obey the rule of the lowest number and the rule of the lowest sum and to be utilized only when the IUPAC nomenclature becomes too difficult.

Examples

[a] 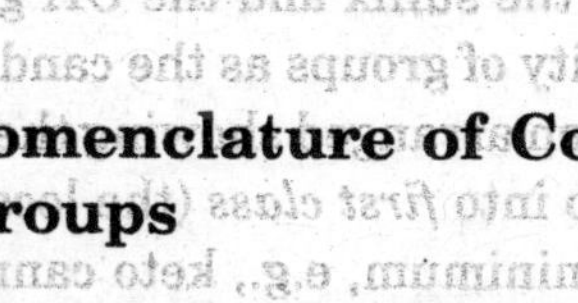

→ 3-Amino-4-chloro-2-hydroxy-1-nitrobutane (locants set from right to left) [A]

→ 2-Amino-1-chloro-3-hydroxy-4-nitrobutane (locants set from left to right) [B]

The alphabetic order has been obeyed. In both [A] and [B] the sum of locants is same, so the rule of the lowest number is to be followed, thus [B] starting with 2 is correct numbering, not [A] which starts with greater number (i.e., 3).

But if it is known that group (as minor suffix 'ol') has the greatest priority over the three other substituents, then the IUPAC name of the compound [C] can be done very easily and this nomenclature is the best. Thus [C] is 2-amino-1-chloro-4-nitrobutan-3-ol.

[b] 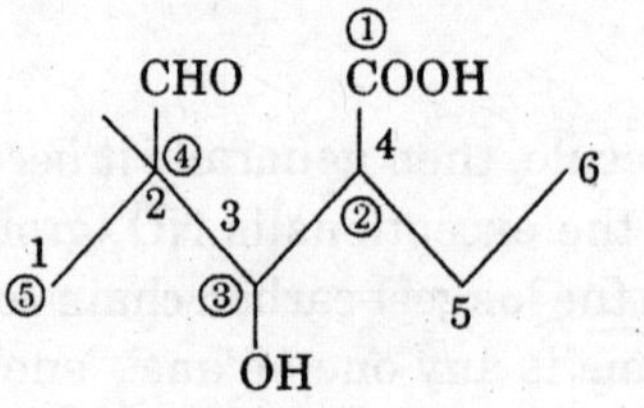

[F] = locant circled

®4-Carboxy-2-formyl-3-hydroxy-2methylhexane
[as a derivative of hexane]
[D]
or
3-Carboxy-5-formyl-4-hydroxy-5-methylhexane
[E]

The sum of locants in [D] is (4+2+3+2=11) and that of [E] is (3+5+4+5=17). Therefore, [D] is acceptable by the rule of lowest sum (here '4s' is not applicable since the 'sums' are not equal in [D] and [E].

But it is known that carboxy group always becomes the minor suffix and its carbon is always marked as 1 (i.e., the numbering of the longest chain starts from the carbon of the COOH group) and following only the alphabetic order of the substituents. Since the carboxy group always holds the locant 1, this locant of the carboxy group is not mentioned in the IUPAC nomenclature [F], which becomes 2-ethyl-4-formyl-3-hydroxy-4-methylpentanoic acid.

It is to be noted that in the first example [a], the OH group became the suffix as 'ol' and in the second example [b], the COOH group becomes the suffix and the OH group becomes prefix as 'hydroxy'. So there is an order of priority of groups as the candidate for the minor suffix. Here these functional groups have been arranged showing the order of priority with an arbitrary classification of these group into *first class* (the locant is always 1), *second class* (the locant may or may not be minimum, e.g., keto cannot have locant 1, in aliphatic compound, but an aldehyde can) and the *third class* (these are mostly used as prefixes and hardly as the suffixes). These are presented in Sec. 63.

63. A. Mnemonic for Priority Order of Functional Groups

This has been explained in the following chart. The following mnemonic may be useful to remember the priority order. Thus,

KOO >	KOOR >	KOK >	KON >	Shall >	allure >	cynic >	cats;
↓	↓	↓	↓	↓	↓	↓	↓
COOH gr. (Acid)	CO_2R (Acid)	COCl (Acid chloride)	$CONR_2$ (Acid amide)	C-SO_3H (Sulphonic acid)	CHO (Aldehyde)	CN (Cyano compounds)	>=0 (Ketones)

> OH >	Am >	I >	Energetic? ; >	O >	Floor >	Clean >	Bright >	Ionic
↓	↓	↓	↓	↓	↓	↓	↓	↓
OH gr. (Alcohol)	NR_2 (Amine)	yne (C≡C)	ene (C=C)	C–O–C (Ether)	F (Fluorine)	Cl (Chlorine)	Br (Bromine)	I (Iodine)

Chart [I]

[i] In this chart the upper functional group will be suffix and the lower will be prefix.

[ii] Among many functional groups, only one will be suffix (greatest priority, i.e., uppermost) and other will be prefixes. The major suffix (ane, ene, etc.) must stay.

[iii] NO_2 group is always prefix.

B. Various Examples of Applications of the Mnemonics

The following abbreviation have been used here; Pr. = Prefix; min.s. = minor suffix, * = locants necessary due to existence of other isomeric compounds with the same carbon chain, f = locants unnecessary (vide 20B).

[I] (a) *Halogen compounds*: Halo (Pr.), halide (min.s.), e.g., haloalkane (single word) or alkyl halide (two words)

Classes and degrees :

[i] *Primary*, (halogen is connected to a *primary* carbon atom)

e.g., $R—CH_2—Cl$, marked as 1° (i.e., *primary* chloro-compound).

$R—CH_2—X$, 1°-(*primary* halocompounds)

↑ 1°C

$CH_3—X$ = 1°-halo-compounds (Methyl halide).

[ii] *Secondary* halo compounds: (halogen is connected to the 2° carbon)

e.g., $R(R_1)$>CH—Cl = 2°–chloro-compounds (or *secondary* chloro-compounds)

$R—CH(X)—R_1$ = 2°–halo-compounds.

[iii] *Tartiary* halo-compounds (halogen is connected to the 3° carbon).

e.g., $R(R_1)(R_2)$>C—Cl = 3°-chloro-compound (or *tertiary*chloro-compounds).

$R(R_1)(R_2)$>C—X = 3°-halo-compounds (or *tertiary*halo-compounds)

[iv] Any combination of 1°, 2° and 3° halo-compounds

Example: $R—CH(X)—C(R_1)(X)—CH_2X$ (X may be any combination of F, Cl, Br and I), with the CH marked 2°, the $C(R_1)$ marked 3° and the CH_2X marked 1°.

$R \neq H, R_1 \neq H$.

It is to be noted that designation 1°, 2° and 3° of halo-compounds depends on the designation of the carbon to which the halogen is attached as a ligand. (vide alcohols and amines). Also halogens directly attached to the aromatic nucleus do not classify the compound as 1°, 2° and 3°, but they are called aromatic halo compounds and their chemical properties differ from those of 3° halo compounds although these two classes appear to be related to each other by similarity of the connecting carbons.

Chart [I] **Upper class** (tends to become suffixes)

First Class functional groups. The C holding the group is always marked as C_1

- COOH (Carboxy; oic)
- ↓ –COOR (oate, ester)
- ↓ –COCl (acid chloride)
- ↓ $-CONR_2$ (acid amide)
- ↓ $-C-SO_3H$ (Sulphonic acid)
- ↓ –CHO (oxo, al, aldehyde)
- ↓ –CN (cyano, nitrile)

Never C_1 as locant (in aliphatic compd.)

- ↓ $\underset{C}{\overset{C}{}}\!>\!C{=}O$ (Keto, oxo, one) (minimum locants is 2 in aliphatic compd.)

2nd Class F.G. may or may not be C_1.

- ↓ –OH (hydroxy; ol)
- ↓ $-NR_2$ (amino, amine)
- ↓ –C≡C (yne)
- ↓ –C=C (ene)

3rd Class F.G. most of the cases these are not suffix but become prefix.

- ↓ C—O—C (alkoxy, ether oxa)
- ↓ F (fluorine fluoro, fluoride)
- ↓ Cl (chlorine, chloro, chloride)
- ↓ Br (bromine, bromo, bromide)
- ↓ I (iodine, iodo, iodide)

Examples of application (Suf.=Suffix; Pr=Prefix)

[i] $OHC-\overset{1}{C}OOH$ = –COOH (Suf), –CHO (Pr.) (COOH Vs. CHO) –CHO (Pr.)

(Oxoethanoic acid)

[ii] $CH_3-CO-CH_2-\overset{1}{C}HO$ = –CHO (Suf.) (CHO Vs. Keto) >=O (Pr.)

(3-Ketobutanal)

[iii] $HO-CH_2-CO-CH_2-\overset{1}{C}HO$ = –CHO (Suf.) >=O (Pr.) –OH (Pr.)

(4-Hydroxy-3-ketobutanal)
H>K—an (major)—al (minor) Suf.

[iv] $HO-CH_2-\underset{\underset{O}{||}}{C}-C\equiv CH$ >C=O (Suf.) HO; (Pr.)

C≡C (Pr.) (major Suf.)
(1-Hydroxybut-3-ynone)

[v] $Cl-CH_2-\underset{\underset{OH}{|}}{CH}-CH=CH_2$ –OH (Suf.), C=C (major Suf.) Cl (Pr.)

(1-Chlorobut-3-en-2-ol)

[vi] (a) $CH_2=CH-C\equiv CH$ C≡C (Suf.), yne Vs. ene C=C (Pr.)

(Butenyne)

(b) $\overset{1}{C}H_2=CH-CH_2-C\equiv\overset{1}{C}H$ (Pent-1-en-4-yne) (4S)

[vii] $CH_3-\overset{4}{O}-CH_2-CH=\overset{1}{C}H_2$ C=C (Suf.), C—O—C (Pr.)

4-oxapent-1-ene

[viii] $O_2N-CH_2-\underset{\underset{I}{|}}{CH}-CH_3$ 3rd Class functional group

(2-Iodo-1-nitropropane); ane (major suffix); NO_2 (Pr.), I (Pr.)
I > n (alphabetic order)

Lower class (tends to become prefixes)

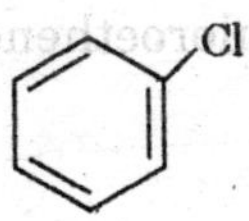 = Aryl chloro– $(CH_3)_3C—Cl$ 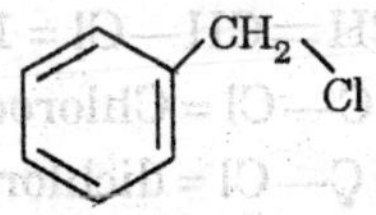 $\equiv$ Ar—CH_2—X (1°C)

Aromatic chloro-compound (Chlorobenzene); 3°-chloro-compound (*tert*-butyl chloride) (or 2-chloromethylpropane)(two words) (single word); 1°-chloro-compound (Benzyl chloride); X=halogen (Cl)

[v] **Classification on the basis of the parent compounds:**

Groups: Allylic (CH_2=CH—CH_2—), e.g., CH_2=CH—CH_2—Br = allyl bromide

e.g., *CH_3—CH=CH—CH(CH_3)—Br = 2-Bromopent-3-ene

e.g., * 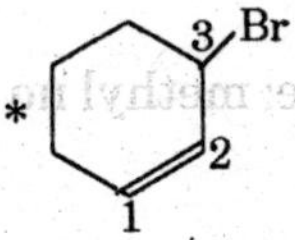 **= 3-Bromocyclohexene. (no locant for 'ene').**

*** shows locant(s) necessary**

e.g., * Benzylic: Ph—CH_2—X (1°–), Ph—CH(C)—X (2°–), Ph—C(C)(C)—X (3°–)

e.g., Vinylic : CH_2=CH—X,

e.g., Aryl: 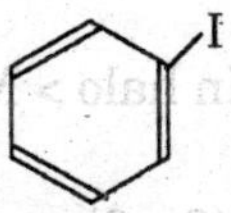 **(Iodobenzene)** 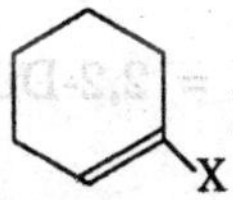 **I-Halocyclohexene**

Me(CH_2)$_2$—CH_2—X = (*n*-alkyl halide); Me_2CH—X = isoalkyl halide (2°), Me_2CH—CH_2—X = isoalkyl halide (1°);

Me_3C—CH(Me)—X = *neo*alkyl halide (2°), Me_3C—CH_2X = *neo*alkyl halide (1°),

Me_3C—C(Me)$_2$—X = *neo* alkyl halide (3°), 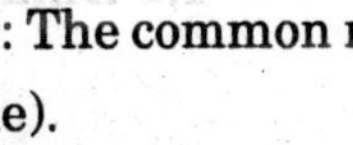 **= *sec*-alkyl halide (2°)**

[b] *Common and IUPAC names of alkyl halides*: The common names are in parenthesis.

1. **MeCl= Chloromethane (methyl chloride).**
2. **Me_2CHI = 2-iodopropane (*iso*-propyl iodide).**
3. **CH_2Cl_2 = Dichloromethane (methylene chloride).**
4. **$CHCl_3$ = Trichloromethane (chloroform).**
5. **CCl_4 = Carbon tetrachloride (Note: nomenclature for binary compounds).**
6. **Cl_2CH—CH_2—Cl=1,1,2-trichloroethane, Cl_3C—CH_3 = 1,1,1-trichloroethane.**
7. **CH_2=CH—Cl = Chloroethene (vinyl chloride)**

8. Cl—CH=CH—Cl = 1,2-dichloroethene; CH_2=CCl_2 = 1,1-dichloroethene
9. CH≡C—Cl = Chloroethyne
10. ClC≡Ç—Cl = dichloroethyne
11. CH_3—$CHBr_2$ = Ethylidene dibromide (*gem*-dibromoethane; halogens on the *same* carbon)
12. CH_3—CBr_2—CH_3 = *iso* propylidene dibromide (*gem*–)
13. CH_3—CHBr—CH_2—Br = 1,2-dibromopropane (*vic*-dibromopropane)

[c] *IUPAC nomenclature of organic halo-compounds*:

1. $BrCH_2$—CH_2—Br (Ethylene dibromide or *vic*-dibromoethane) = 1,2-Dibromoethane.
2. CH_3—$CHBr_2$ = 1,1-Dibromoethane [cf. example 11 in (b) above].
3. CH_3—CH(CH_3)—CH_2(—Cl) = 1-Chloromethylpropane (Note: methyl no locant).
4. CH_3—CCl_2—CH_3 = 2,2-Dichloropropane
5. CH_3—C$(CH_3)_2$—CH_2—Cl = Chlorodimethylpropane (no locant)
6. CH_3—CMe_2—Cl = 2-Chloromethylpropane = (Me_3C—Cl) (*t*-Butyl chloride)
7. 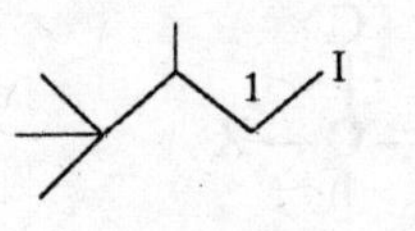 = 1-Iodo-2,3,3-trimethylbutane (sum of locant 9)

   not 4-Iod-2,2,3-trimethylbutane (do = 11)
   (Rule of lowest sum)
8. 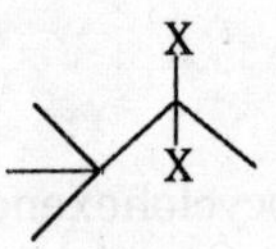 = 2,2-Dihalodimethylbutane (H in halo > M in methyl);

   not 3,3-Dihalodimethylbutane (2 < 3)
   (Rule of lowest sum).
9. 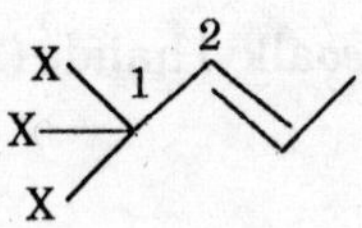 = 1,1,1-Trihalobut-2-ene (but+ene) (a in buta is dropped)
10. 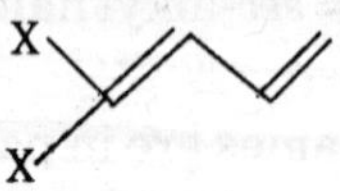 = 1,1-Dihalobuta-1,3-diene (a stays, followed by d in di)
11. 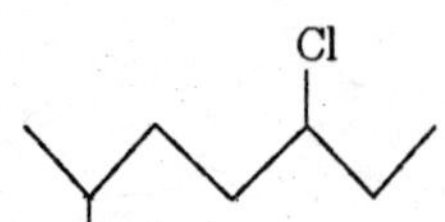 = 5-Chloro-2-methylheptane; (2+6 = 8 = sum of locants)

    not 3-Chloro-6-methylheptane (3+6 = 9 = sum of locants)
    (Rule of lowest sum > Rule of lowest number)

12. 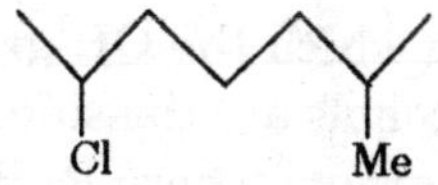 = 2-Chloro-6-methylheptane (2+6 = 8 = sum of locants)

not, 6-Chloro-2-methylheptane (2+6 = 8 = sum of locants)

13. $ClF_2C—CF_2Cl$ = 1,2-Dichlorotetrafluoroethane (no locant of F)

14. $Cl_2FC—CFCl_2$ = Tetrachloro-1,2-difluoroethane (locant of F needed)

15. $n–C_5H_{11}–\overset{5}{C}H–(CH_2)_2–\overset{2}{C}H–Me$ (C-5 bears Cl–CH–CH$_2$–Me; C-2 bears Br) = 2-Bromo-5-(1-chloropropyl)–decane

16. $n–C_5H_{11}–\overset{5}{C}H–(CH_2)_2–\overset{2}{C}H(Br)–Me$ (C-5 bears $\underset{1}{CH_2}—\underset{2}{CH_2}—\underset{3}{CH_2}Cl$) = 2-Bromo-5-(3-chloropropyl)–decane

17. $F_3C—CF_2—CF_2—CF_3$, perfluorobutane (all H replaced by F, so 'per' is added to the halogen).

It is to be noted that simple multiplying prefix (mono, di, tri, tetra, etc.) is only a multiplying prefix and not prefix itself. So the alphabetic order rule is not applicable to the multiplying prefix [cf., e:g., 13 and 14 C(in chloro) come before F(in fluoro)] but T (in tetra) comes before D in di] because these are only simple multiplying prefixes. Complex multiplying prefix [e.g., *bis*, *tris*, *tetra*, etc.] includes its first letter during selection of alphabetic order.

[II] *Nomenclature of alcohols*

[a] The C—OH bonding is characteristic of generating alcholos and this C—OH must not be bonded to any heteroatom on this carbon. Thus —C(=O)—OH, —C(O—R)—OH, —C(NH$_2$)—OH, etc. are not to be considered as alcohols. The prefix of the –OH group is 'hydroxy', the suffix is 'ol', and the name is hydroxyl. The major suffix may be ane, ene, yne and enyne, the minor suffix is ol, and the last vowel of the major suffix is cancelled. The low-molecular weight alcohols are more commonly known by their trivial names but the higher molecular weight members are better known by IUPAC names, specially when substituents are present in the alcohol molecules.

Examples

= Butan-1-ol or 1-Butanol (it is better to put the locants just before each substituent). This is Butane + ol ('e' cancelled) = Butanol.

[b] *Classification of alcohols*: The degree of carbon to which the OH group is attached determines the class. Thus 1°, 2° and 3° alcohols are classified.

[i] CH_3—OH = 1° alcohol (special case of RCH_2-group, where R=H); methanol; methyl alcohol; logic, methanol behaves like other conservative 1°-alcohols like ethanol, propan-1-ol in most of the chemical reactions, e.g.,

Primary Alcohol

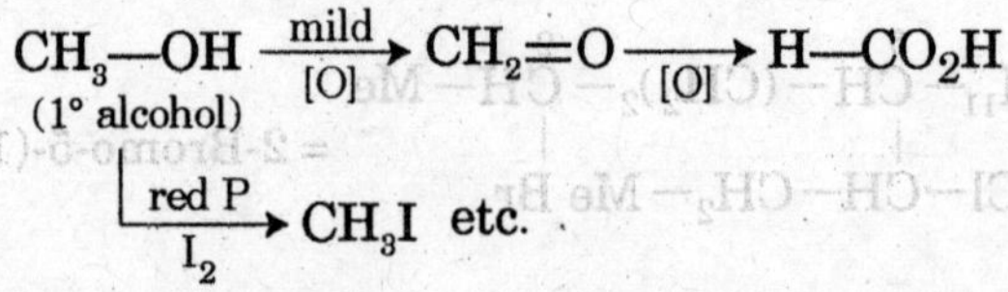

Other primary alcohols respond to all these tests but formaldehyde and formic acid are both reducing agents (difference).

CH_3—CH_2—OH = Et—OH = R—CH_2—OH (R=CH_3)

Primary Alcohol = Ethanol; Ethyl Alcohol

[ii] $(CH_3)(CH_3)$>CH—OH = Propan-2-ol

(Secondary alcohol) (2° alcohol) (*iso*-propanol)

CH_3—CH_2—CH_2—OH (Propan-1-ol), isomeric,

(*n*-propanol)

[iii] 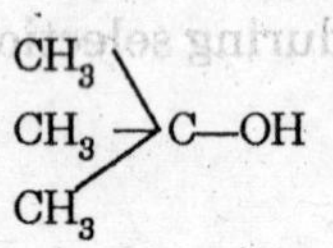 = Methylpropan-2-ol,

(*tert*-alcohol) (3°) (3° alcohol) and (*tert*-butanol or *t*-butanol)

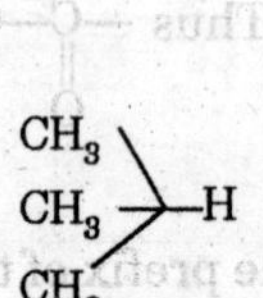 = Methylpropan-1-ol (isomeric)

(1° alcohol)

[iv] Another class is *neo*-alcohols which may be 1°, 2° or 3° but a *neo*-carbon is present (*neo*-carbon is a carbon connected to four other carbons directly). Thus,

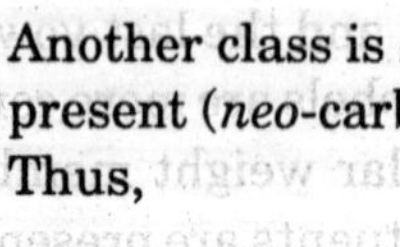 = Dimethylpropanol (1°-alcohol)

(*neo*-pentyl alcohol) (five carbons, so pentyl)

[v] When an aromatic group is connected to the carbon holding the OH group, directly or indirectly like $Ar(C)_n$—OH (Ar = Aromatic rings, $n \neq 0$) aromatic alcohol (i.e., aryl alcohols) are formed (but $n = 0$ generates phenols, which have different chemistry and they are not under alcohol although they have minor suffixes 'ol'.

[vi] Classification on the basis of the number of hydroxyl group: These are (a) monohydric (contains one OH gr.) alcohols, e.g., R—OH, EtOH; (b) dihydric, e.g., $R(OH)_2$, HO—CH_2—CH_2—OH; (c) trihydric, e.g., $R(OH)_3$, $HOCH_2$—CH(OH)—CH_2OH, (d) polyhydric, e.g., $R(OH)_x$; and (e) *geminal* diol, the two OH groups on the same carbon, e.g., CH_3—$CH(OH)_2$, ethane-1,1-diol; (f) *vicinal* diol, the two –OH groups on adjacent carbons, e.g., $HOCH_2$—CH_2—OH, ethane-1,2-diol (geminal is often abbreviated as *gem* and vicinal as *vic*).

[c] *Examples of alkanols*: Various examples are given below,

[i] Butanols ($C_4H_{10}O$, isomers): CH_3—CH_2—CH_2—CH_2—OH, *n*-butyl alcohol; (Butan-1-ol in IUPAC), also *sec*butyl, *iso*butyl and *tert*butyl alcohols are named in trivial system, but in IUPAC system the root name changes for *iso*butyl and *tert*butyl isomers (II b).

[ii] Butanediols ($C_4H_{10}O_2$, Isomers):

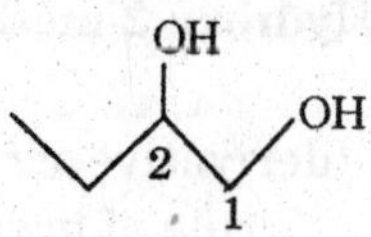

(Butane-1,2-diol)
vicinaldiol
i.e., *vic*-diol

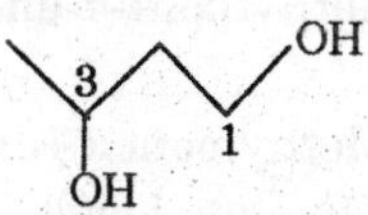

(Butane-1,3-diol)

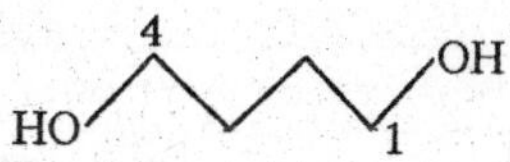

(Butane-1,4-diol)

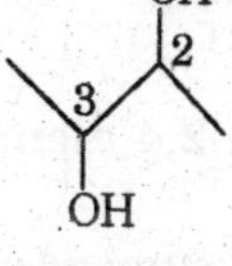

(Butane-2,3-diol)
i.e., *vic*-diol

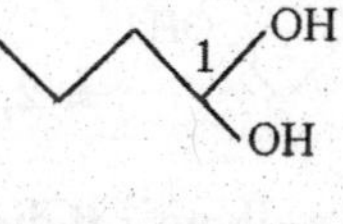

(Butane-1,1-diol)
geminal diol
i.e., *gem*.diol

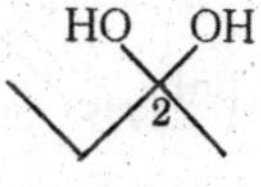

(Butane-2,2-diol)
(Note: butane, not butan)

[iii] 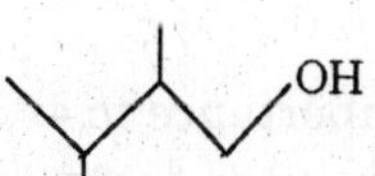 = 2,3-Dimethylbutan-1-ol.

[iv] 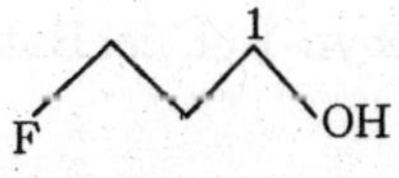 = 3-Fluoropropan-1-ol.

[v]  = 2-Methylbutan-3-ol (not 3-methylbutan-2-ol, cf. 4s)

[vi] 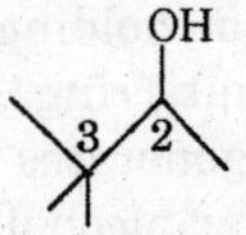 = 3,3,-Dimethylbutan-2-ol (4s not applicable, so minor suffix gets lower locant ignoring the rule of lowest sum).

[vii] $Cl_3C—CH(OH)_2$ = 1,1,1-Trichloroethanediol (no need of locant of 'ol')

[d] *Examples of alkenols*: These are unsaturated alcohols with D.U. 1 (i.e. one C=C)

[i] 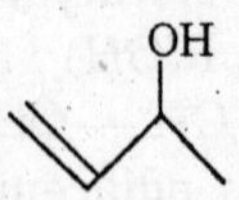 = But-3-en-2-ol (But~~a~~+en~~e~~+ol; 'ol' gets lower locant)

[ii] 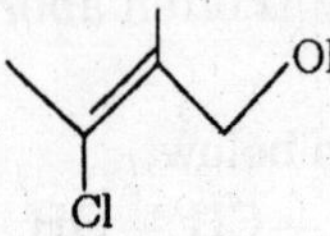 = 3-Chloro-2-methylbut-2-en-1-ol. (Alphabetic order)

[iii]  = Prop-2-en-1-ol (Allylic alcohol)

[iv] 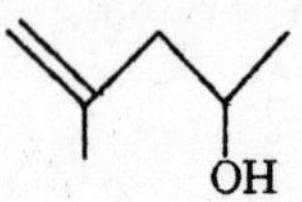 = 4-Methylpent-4-en-2-ol or 4-Hydroxy-2-methylpent-1-ene,

(4-Methylpent~~a~~+en~~e~~+ol • ol as minor suffix gets lower locant) (derivative of hydrocarbon, rule of lowest sum)

[v]  = 5-Methylhex-4-en-1-ol

(ol as minor suffix gets lower locant)

[vi] 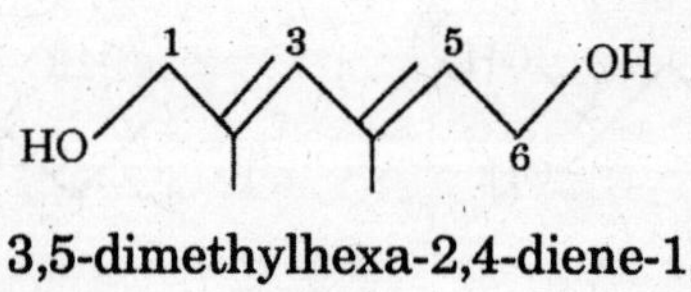 = 2,4-Dimethylhexa-2,4-diene-1,6-diol. Not, 3,5-dimethylhexa-2,4-diene-1,6-diol. (Rule of lowest sum; no elision of vowel).

[e] *Examples of alkynols*: These are unsaturated alcohols with D.U. 2 (i.e., one C≡C)

[i] $HC{\equiv}C—CH_2OH$ = Prop-2-yn-1-ol (given lower locant to ol)
(Propargyl alcohol) = Prop~~a~~+yn~~e~~+ol
Propiolic alcohol or Prop-1-yn-2-ol (given importance to 4s, but the other name is used to give priority to ol)

[ii] 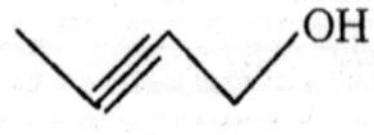 = But-2-yn-1-ol (*not* Butan-2-yn-1-ol, *not* Buta-2-yn-1-ol)

[iii] 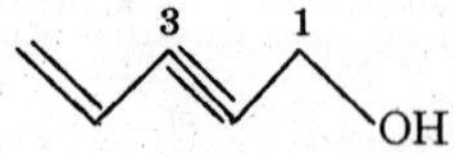 = Pent-4-en-2-yn-1-ol (alphabetic order e>y)

Note: There was another system of naming the alcohols as the derivatives of 'carbinols' which have the characteristic [C—OH] group. Thus

[i] 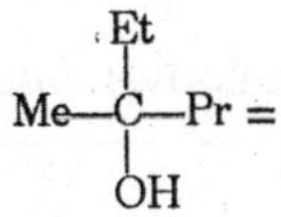 = Ethylmethylpropyl carbinol (alphabetic order e > m > p)

[ii] $Me_3C—CH_2OH$ = *tert*.Butyl carbinol

[III] *Nomenclature of ethers*: The general formula of ether is $R—O—R_1$ ($R = R_1$ gives *symmetrical* ether and $R \neq R_1$ gives *unsymmetrical* ether), Ar—OR (Ar = Aromatic, R = aliphatic gives aryl alkyl ether) and Ar—OAr (Ar = Ar′ gives *symmetrical diaryl* ether and Ar≠Ar′ gives unsymmetrical diaryl ether). The –OR group is called alkoxy group. (e.g., $–OCH_3$ as methoxy, $–OC_2H_5$ as ethoxy, etc).

[a] [i] The *symmetrical dialkyl ethers* are generally named as such (i.e., naming the alkyl group with the multiplying prefix 'di', then adding 'ether' as separate word. Thus $C_2H_5—O—C_2H_5$ (or Et_2O) is called diethyl ether; $(CH_3)_2CH—O—CH(CH_3)_2$ as diisopropyl ether, etc.

[ii] The *unsymmetrical dialkyl ethers* are named in two ways. The first method is to name the alkyl groups in alphabetic order as separate words, then adding the ethers as end word. Thus, $C_2H_5—O—CH_3$ is named ethyl methyl ether, $C_3H_7^n—O—CH_3$ is named as methyl *n*-propyl ether etc. The second method is to name the smaller alkyl group as *alkoxy* and the larger alkyl group as hydrocarbon. Thus, $C_2H_5—O—CH_3$ is methoxyethane and $C_3H_7^n—O—C_2H_5$ is ethoxypropane, here the alkoxy group is a substituent of the hydrocarbon. So in R—O—R′ if R is smaller than R′, the R is alkoxy and R′ is the alkane.

[iii] In aromatic ether when Ar=C_6H_5gr (i.e., Ph gr.), then the ether is a *phenolic ether* (i.e. Ph—O—R) which is named as alkoxybenzene assuming Ph is larger than R. The lower members of this series also have well known common names. Thus Ph—O—CH_3 is methoxybenzene or anisole, Et—O—Ph as ethoxybenzene or phenetole etc. The OPh group is phenoxy.

[iv] In aromatic ether when Ar and Ar′ are equal, then *symmetrical diaryl ether* is formed and named as such. Thus Ph—O—Ph is diphenyl ether, $C_{10}H_7—O—C_{10}H_7$ is named dinaphthyl ether (when $C_{10}H_8$ is naphthalene).

[v] In aromatic ether, when Ar and Ar′ are different, then *unsymmetrical* diaryl ether is formed and named as aryl aryl ether following the alphabetic order of the names of the each aryl group. Thus $C_{10}H_7—O—C_6H_5$ is naphthyl phenyl ether.

[b] A very useful and interesting method for naming the difficult compounds is as follows.

[i] To consider the O as carbon and get the total number of carbon in the largest chain (including the oxygen) to determine the root name.

[ii] To denote ether as 'oxa' and giving the lowest locant to the oxygen when possible, and expressing as 'oxaalkane'.

Examples

$C_3H_7—O—C_2H_5$ = 3-Oxahexane (not 4-oxahexane), *lowest sum*.

$\overset{6}{C}H_3—\overset{5}{C}H_2—\overset{4}{O}—\overset{3}{C}H(CH_3)—\overset{2}{C}H(Cl)—\overset{1}{C}H_3$ = 2-Chloro-3-methyl-4-oxahexane

(Rule of lowest sum)

and $\overset{1}{C}H_2{=}\overset{2}{C}H—\overset{3}{C}H(CH_3)—\overset{4}{O}—\overset{5}{C}H(CH_3)—\overset{6}{C}H_3$ = 3,5-Dimethyl-4-oxahex-1-ene

(Rule of lowest sum).

Also $CH_3—\overset{O}{\overset{\|}{C}}—O—\overset{O}{\overset{\|}{C}}—CH_3$ = 2,4-Diketooxapentane

(Acetic anhydride) (Ethanoic anhydride)

(Note: Oxa is —O— and oxo is $—\underset{\underset{O}{\|}}{C}—$ group)

Special examples:

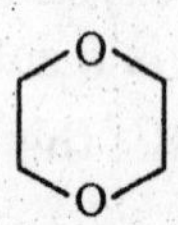 = 1,4-dioxacyclohexane (dioxane) 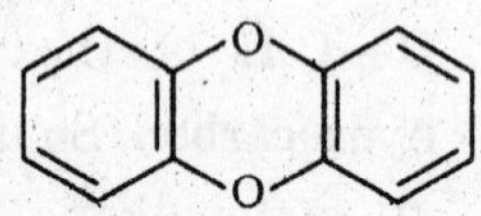 (dioxine).

 = Epoxide, 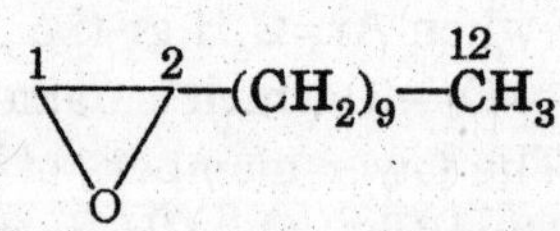 =1,2-Epoxydodecane.

$MeO—CH_2—CH_2—O—(CH_2)_2—OMe$ = Diglyme

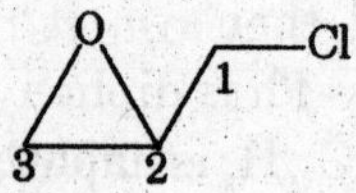 = 1-Chloro-2,3-epoxypropane

[IV] *Nomenclature of Ketones*: The functional group is $C—\overset{O}{\overset{\|}{C}}—C$ named keto as prefix, oxo (*not oxa*) as prefix, and –one (*pr.* own) as minor suffix. The keto group and the formyl group (i.e., aldehyde group, –CHO) are often called *carbonyl* group. When in a molecule both aldehyde and keto groups are present, then aldehyde gets preference and it becomes the minor suffix and keto (i.e., oxo) becomes prefix (vide mnemonic).

Examples

[a] Simple ketone (Alkyl groups are same) $Me_2C=O$, $R_2C=O$, mixed ketone (Alkyl groups are different (MeCO–Et, R—CO—R′)

[b] Common method of marking the carbons (by Greek alphabets),

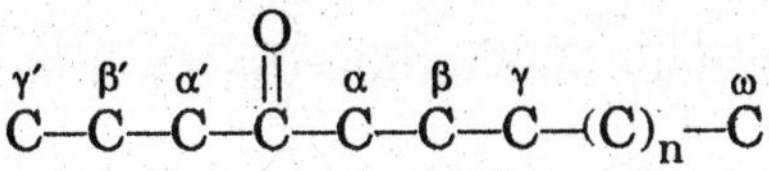

(ω is the last carbon), Thus,

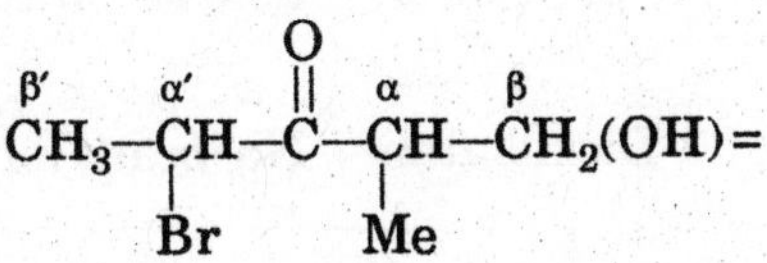

= α-Methyl-α′-bromo-β-hydroxypentanone (α > α′).

[c] 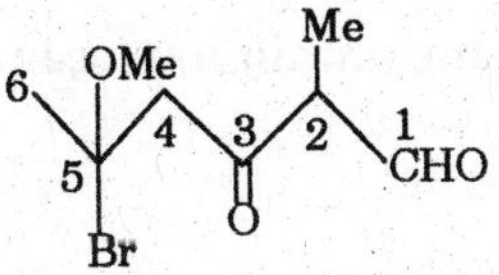

= 5-Bromo-5-methoxy-2-methyl-3-oxohexanal.

B → metho → methy → oxo → minor suffix
(al > oxo)

[d] IUPAC method of nomenclature of aliphatic ketone.

[i] 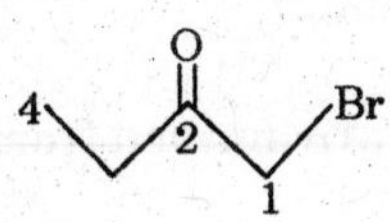

= 1-Bromobutanone;

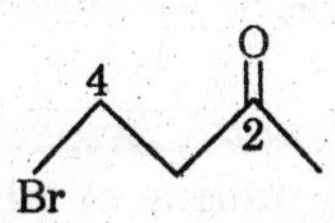

= 4-Bromobutanone

(Butanone is 2-keto)

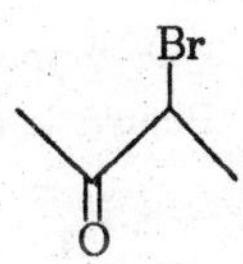

= 2-Bromobutanone (not 3-bromobutanone) (no locant for 'one')

(3 > 2)

[ii] 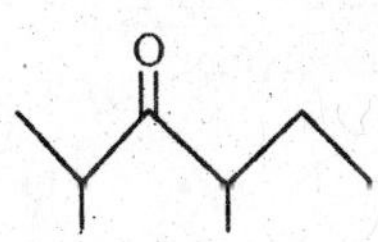

= 2,4-Dimethylhexan-3-one. (Rule of lowest Sum)

[iii] 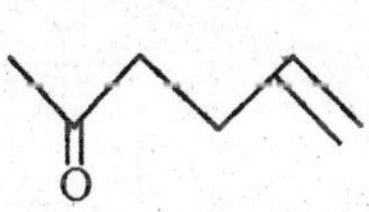

= Hex-5-en-2-one (one > ene) (5-Ketohex-1-ene)

[iv] 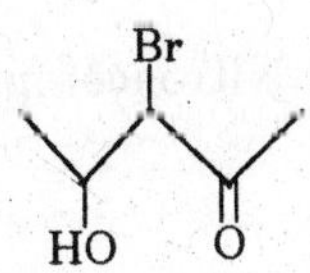

= 3-Bromo-4-hydroxypentanone (one > hydroxy > bromo)

B ———→ H ———————→ one (minor suffix)
(only one position is available for 'one')

[v]  = 4-Ketopentanal (4-oxopentanal)

[vi] 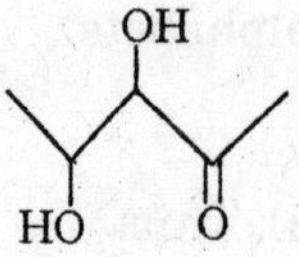 = 3,4-Dihydroxypentanone (one retains the lower locant which is omitted here because C_2 is the only position available for 'one').

[vii] 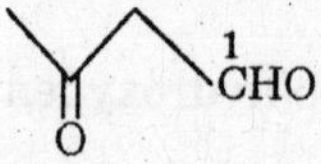 = 3-Oxobutanal (or 1,3-Dioxobutane). No locant for 'al'.

[viii] 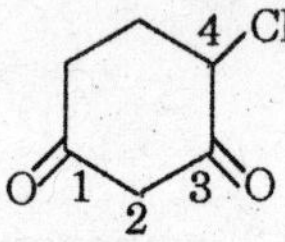 = 4-Methylcyclohexane-1,3-dione (minimum locants for one)

[ix] 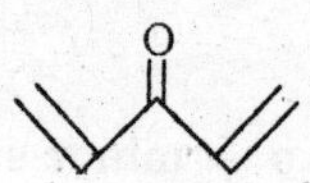 = Penta-1,4-dienone (common name divinyl ketone).

[x] 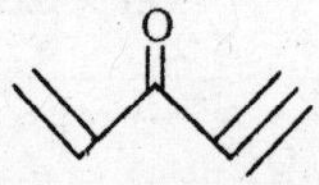 = Pentenynone (Pent-1-en-4-ynone)

[e] *Naming of aromatic ketones*: The common names are derived from the aliphatic acids with the ending of 'phenone'. Thus,

[i] $CH_3—CO—OH \xrightarrow{-[-OH]} CH_3—CO— \longrightarrow CH_3—CO—C_6H_5$

acetic acid (aceto–) acetophenone

(Phenyl ketone = Phenone)

[ii] $CH_3—CH_2—COOH \longrightarrow CH_3—CH_2—CO– \longrightarrow CH_3—CH_2—CO—Ph$

propionic acid (propio) propiophenone

[iii] $C_6H_5CO—OH \longrightarrow C_6H_5—CO– \longrightarrow C_6H_5—CO—C_6H_5$

benzoic acid (benzo-) benzophenone

[iv] 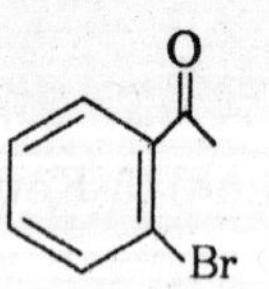

ortho-Bromoacetophenone

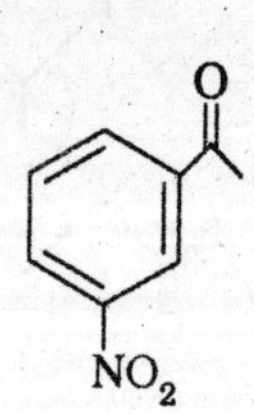

meta-Nitroacetophenone

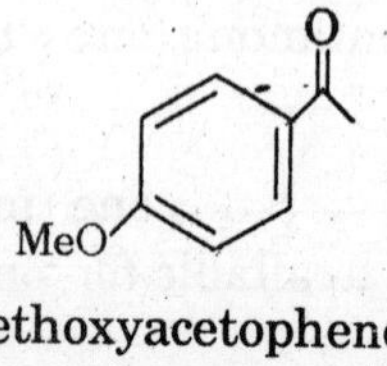

para-Methoxyacetophenone

[f] Various examples with systematic nomenclature.

[i] $CH_3—\overset{1}{C}O—C_6H_5$ = (acetophenone) = 1-Oxo-1-phenylethane

[ii] $\overset{3}{C}H_3—\overset{2}{C}H_2—\underset{1}{\overset{O}{\overset{||}{C}}}—C_6H_5$ = (propiophenone) = 1-Oxo-1-phenylpropane

[iii] $CH_3—CH=CH—CO—CH_3$ = Pent-3-enone

[iv] $CH_3—CO—CH(CH_3)_2$ = Methylbutanone (no locant)

[v] $CH_3—CO—CMe_3$ = Dimethylbutanone (no locant)

[vi] $Me—CH_2—CO—CH(CH_3)_2$ = Methylpentan-3-one

[vii] $Me—\underset{Me}{\underset{|}{C}H}—CO—\underset{Me}{\underset{|}{C}H}—Me$ = 2,4-Dimethylpentanone

[viii] $Me—CH_2—CO—CMe_3$ = 2,2-Dimethylpentan-3-one

[ix] MeCO—CO—Me = Butanedione (common name diacetyl)

[x] $\begin{matrix} Me \\ Me \end{matrix}\!\!>\!C=CH—CO—Me$ = 4-Methylpent-3-enone
(common name mesityl oxide)

[xi] $CH_2=C=O$ = Oxoethene (common name ketene).

[xii] $CH_2=CH—\underset{O}{\underset{||}{C}}—CH_2—CH_3$ = Pent-1-en-3-one;

[xiii] $CH_2=CH—CH_2—\underset{O}{\underset{||}{C}}—CH_3$ = Pent-4-en-2-one *or*

Pent-1-en-4-one (Rule of lowest sum)

[V] *Nomenclature of aldehydes*: The –CHO group is the functional group for aldehydes, the carbon is marked as C_1, if it is the minor suffix, called carbonyl group, as prefix it is 'oxo' or 'formyl' and as suffix it is 'al' which needs no locant. The root name and the major suffix are determined and 'al' is added to it as suffix followed by marking the whole molecule taking the aldehyde carbon marked as 1. However, the common names of the aldehydes may sometimes use α,β-notations to designate the position of the carbons of the aldehyde molecules. Thus,

[i] $\underset{\delta}{CH_3}—\underset{\gamma}{\overset{Cl}{\overset{|}{C}H}}—\underset{\beta}{\overset{NO_2}{\overset{|}{C}H}}—\underset{\alpha}{\overset{Me}{\overset{|}{C}H}}—\underset{1}{CHO}$ = α-Methyl-β-nitro-γ-chlorovaleraldehyde
(α → β → γ in this order)
(common name)
= 4-Chloro-2-methyl-3-nitropentanal
(IUPAC name)

[ii] $CH_3-\underset{\mid}{\overset{NO_2}{CH}}-\underset{I}{CH}-\overset{1}{C}HO$ = α-Iodo-β-nitrobutyraldehyde (common name)

= 2-Iodo-3-nitrobutanal (IUPAC name)

[iii] $\overset{\beta}{C}H_2=\overset{\alpha}{C}H-CHO$ = Acrolein (Propenal)

[iv] $CH_3-CH=CH-CHO$ = Crotonaldehyde (But-2-enal); [en ¢ +al]

[v] $CH_2=CH-CH_2-CHO$ = But-3-enal.

[a] The CHO group is sometimes named as carbaldehyde (when attached to a ring and after the name of the parent hydrocarbon.)

e.g. C_6H_5CHO = Benzaldehyde (Benzenecarbaldehyde, IUPAC)

[b] IUPAC names (common names):

[i] $CH_3-\underset{_1CHO}{CH}-CH=\underset{Cl}{C}-CH_3$ = 4-Chloro-2-methylpent-3-enal.

[ii] H_2CO = Methanal (formaldehyde)

[iii] CH_3-CHO = Ethanal (acetaldehyde)

[iv] CH_3-CH_2-CHO = Propanal (propionaldehyde)

[v] $CH_3-CH_2-CH_2-CHO$ = Butanal (butyraldehyde)

[vi] $CH_3-(CH_2)_3-CHO$ = Pentanal (valeraldehyde)

[vii]  = Dimethylpropanal (pivalaldehyde, also trimethylacetaldehyde)

[viii] 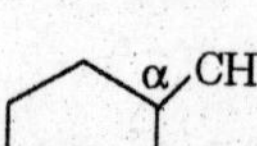 = Methylpropanal (*iso*-butyraldehyde)

[ix] 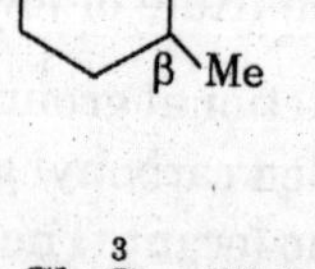 = 1-Formyl-2-methylcyclohexane

(β-methylcyclohexanecarbaldehyde)

[x] $Cl_3\overset{3}{C}-CMe_2-\overset{1}{C}HO$ = 3,3,3-Trichlorodimethylpropanal

[xi] $HO-CH_2-CHO$ = Hydroxyethanal

[xii] $CH_3-CO-CHO$ = 2-Oxopropanal or Ketopropanal

[xiii] $OHC-CH_2-CHO$ = Propanedial.

[xiv] $OHC-CO-CHO$ = Trioxopropane (or Ketopropanedial) [no locant]

[xv] 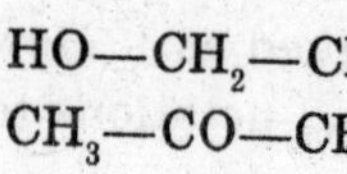 = Methylbutynal; *but*, 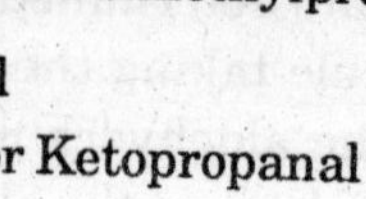 = 2-Methylbut-3-enal,

(no locant)

and 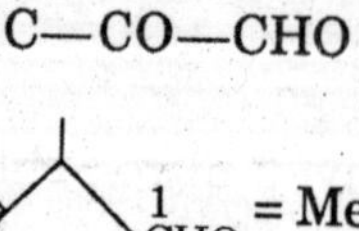 = 3-Methylbut-3-enal. (cf 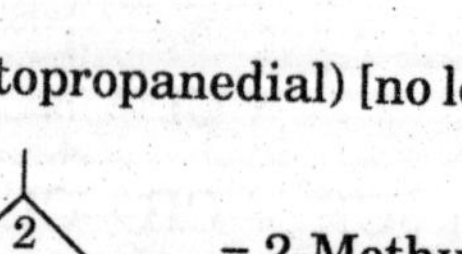)

[xvi] 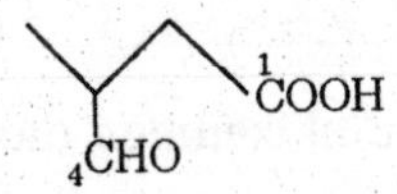 = 3-Methyl-4-oxobutanoic acid (COOH > CHO, mnemonic)

Also, 3-Formylbutanoic acid.

[xvii] 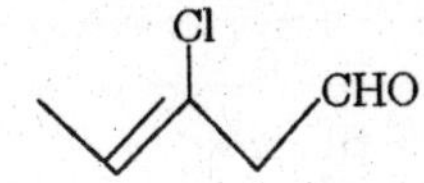 = 3-Chloropent-3-enal.

[xviii] 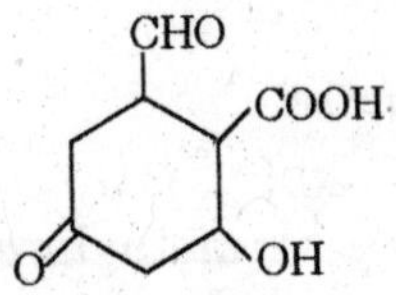 = 2-Formyl-6-hydroxy-4-oxocyclohexanecarboxylic acid

[xix] 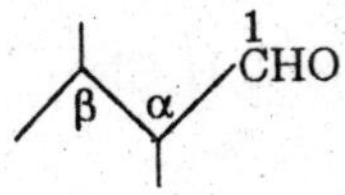 = α,β-Dimethylbutyraldehyde (Geneva system)

= 2,3-Dimethylbutanal (IUPAC system)

[xx] 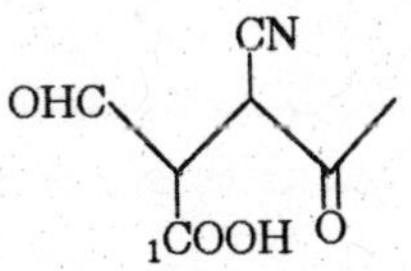 = 3-Cyano-2-formyl-4-ketopentanoic acid

$[C \longrightarrow F \longrightarrow K \longrightarrow C_5\text{—}\overset{1}{C}OOH \text{ gr}]$

[xxi] 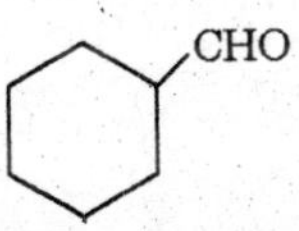 = Cyclohexanecarbaldehyde

[xxii] C_6H_5—$CH(CHO)_2$ = Phenylpropanedial

[**Note:** When used as 'multiplying prefix' penta retains the 'a' before a vowel, e.g., pentaene, which means 5 double bonds, but as 'root name' penta drops 'a' before vowel, e.g., pentenes, i.e., C_5-compound]

[VI] *Nomenclature of cyano compounds*: The cyano group is –CN, the C is marked 1 when used as minor suffix (called nitrile) and when prefix, it is termed cyano. When a cyano group is directly attached to a ring, it is termed as carbonitrile and mentioned at the end. Thus,

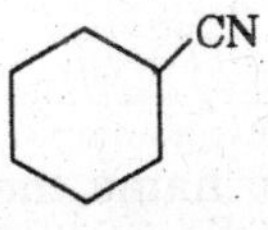 = Cyclohexanecarbonitrile

[a] It is interesting to note that the CN group can form R—CN as well as R—NC. The former class of compounds is called alkyl cyanide and the latter class is called alkyl *iso*-cyanide which is named as carbylamine in the IUPAC nomenclature as shown in parenthesis.

Examples

[i] The *common name* is made by substituting 'ic' of common name of the corresponding acid by 'onitrile'. Thus,

Name of acid in trivial method $\xrightarrow[+(\text{onitrile})]{-(\text{ic})}$ Name of alkyl cyanide

[ii] $CH_3—COOH \longrightarrow CH_3—CN$

Acetic acid → Acetonitrile (ethanenitrile, $\overset{2}{C}H_3 - \overset{1}{C}N$)

[iii] $CH_3—CH_2—COOH \rightarrow CH_3—CH_2—\overset{1}{C}N$

Propionic acid → Propionitrile (Propanenitrile)

[iv] $CH_3—\underset{\large CH_3}{\underset{|}{C}H}—COOH$ → *Iso*butyronitrile; (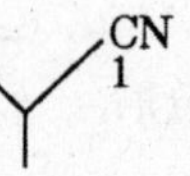 methylpropanenitrile)

[v] $C_6H_5—COOH$ → Benzonitrile (benzenecarbonitrile) C_6H_5CN
(Benzoic Acid)

[vi] $CH_2{=}CH—CN$ → vinyl cyanide; (propenenitrile)

[b] IUPAC method of naming the alkyl cyanides, R—CN

[i] 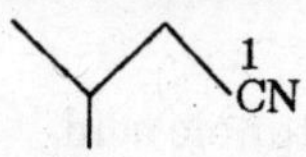 → 3-Methylbutanenitrile

[ii] 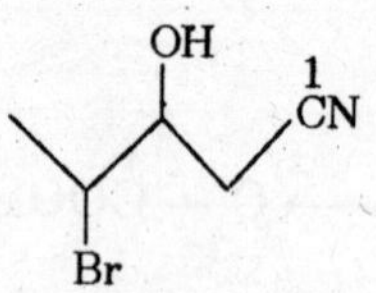 → 4-Bromo-3-hydroxypentanenitrile

[iii] 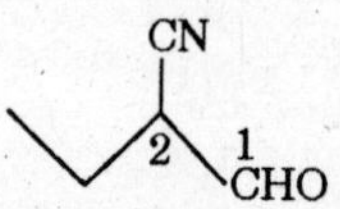 = 2-Cyanobutanal $\left(\overset{1}{C}HO > CN\right)$

[iv] 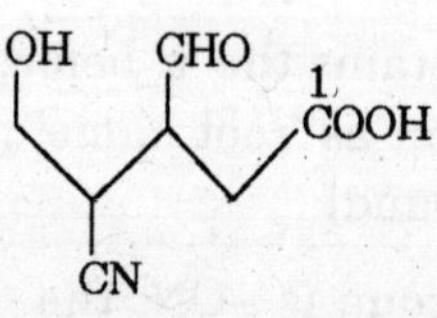 = 4-Cyano-3-formyl-5-hydroxypentanoic acid

(Rule of lowest sum and rule of lowest number are not applicable here as the COOH group must hold the locant 1).

[v] $N\overset{1}{C}—\overset{2}{C}N$ = Ethanedinitrile

[vi] 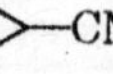 = Cyclopropanecarbonitrile.

[c] *Nomenclature of alkyl isocyanide*: The common method is to name the alkyl group and add the isocyanide as separate word. The IUPAC employ 'carbylamine' after the name of the alkyl group.

Examples

[i] $CH_3—CH_2—NC$ = Ethylcarbylamine (ethyl isocyonide, common name)

[ii] $C_6H_5—NC$ = Phenylcarbylamine (phenyl isocyanide)

[VII] *Nomenclature of amines*: Amines are classified as 1°, 2°, 3° and quaternary salt where the nitrogen of the amino group determines the degree. Another way of classification is the nature of the parent compound which has the amino group as substitution. Thus,

[a] C—NH_2 (R/Ar—NH_2); C—NH—C; C—N(C)—C; $C_4\overset{\oplus}{N}.\overset{\ominus}{X}$

1°-amine; 2°-amime; 3°-amine; Quaternary ammonium salt

[b] R—NH_2; Ar—NH_2; Ar$(CH_2)_n$—NH_2

aliphatic 1°-amine; aromatic 1°-amine; arylalkyl ($n \neq o$) 1°-amine

[c] Amines are named as a single word: Thus, common name and (IUPAC name).

[i] $EtNH_2$ = Ethylamine or Aminoethane (ethanamine; R—NH_2 = alkanamine).
1°-amine

[ii] Et—NH—Me = N-Methylaminoethane (N-methylethanamine)
2°-amino (smaller group is prefixes)

[d] IUPAC naming also permits 'azane' ending,
e.g., $EtNH_2$ = Ethylazane ; Ph_2NH = Diphenylazane,
MeN (Et)Pr = Ethylmethylpropylazane

[e] Other examples:

[i] $(CH_3)_2NH$ = Dimethylamine; [ii] CH_3—NMe_2 = Trimethylamine

[iii] $(CH_3)_4\overset{\oplus}{N}\overset{\ominus}{I}$ = Tetramethylammonium iodide

[iv] Ph—CH_2—$\overset{\oplus}{N}Me_3\overset{\ominus}{I}$ = Benzyltrimethylammonium iodide

[v] MeO—CH_2—CH_2—NHMe = 4-Methoxy-2-azabutane

[vi] CH_3—CH_2—C(=O)—N(Me)Ph = 2-Phenyl-2-azapentan-3-one

or 3-Keto-2-phenyl-2-azapentane

[vii] Ph — $\overset{\beta}{C}H_2$ — $\overset{\alpha}{C}H_2$ — NH_2 = β-Phenylethylamine (common name)
= (2-Phenylethanamine)

[viii] 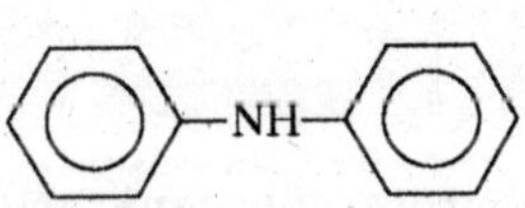 = Diphenylamine; (N-Phenylbenzenamine)

[VIII] *Nomenclature of nitro compounds*: The functional group is –NO_2. The classification is based on the nature (i.e., degree) of the carbon holding the –NO_2 group. Thus,

[a] C—CH_2NO_2 (1°-nitro compound); C—CH(C)NO_2 (2°–); C—C(C)(C)—NO_2 (3°–)

[b] When $-NO_2$ is connected to the alkyl group, then the compound is aliphatic nitro compound, and when $-NO_2$ group is *directly* attached to the aromatic ring, then it is called aromatic nitro compound.
R—NO_2 (aliphatic nitro compound), Ar—NO_2 (aromatic nitro compound).

[c] The nitro compounds are isomeric with the nitrite compounds.
R—NO_2 (nitro alkane) and R—O—N=O (alkyl nitrite).

[d] The nitro group is always used as a prefix, e.g., nitro alkane, and nitroarenes, (i.e., CH_3—CH_2—NO_2 = nitroethane, and $C_6H_5NO_2$ = nitrobenzene).

Examples

[i] Me—CH(Ph)—NO_2 = α-Phenylnitroethane, common name = (1-Phenylnitroethane)

[ii] 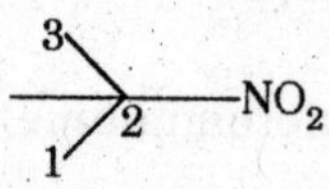 = Methyl-2-nitropropane

[iii] 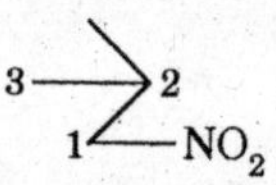 = Methyl-1-nitropropane

[iv] 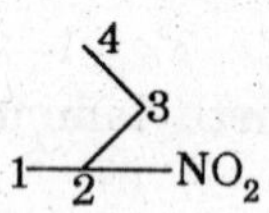 = 2-Nitrobutane

[v] 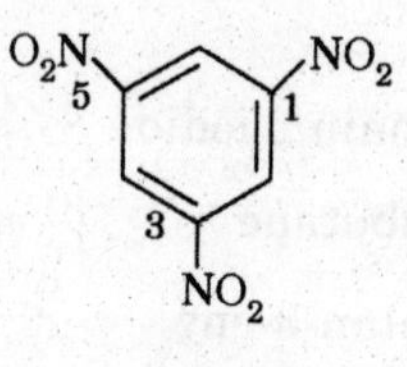 = 1,3,5-Trinitrobenzene (TNB)

[vi] 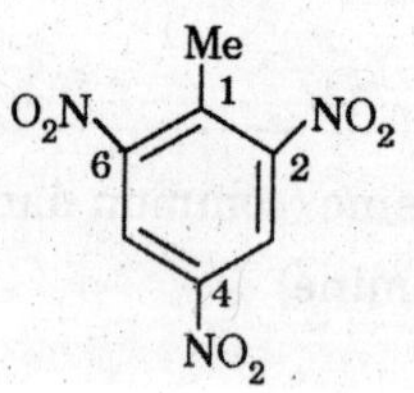 = 2,4,6-Trinitrotoluene (TNT)

[IX] *Nomenclature of carboxylic acids*: The functional group of organic acids is –COOH (The C is always marked as C_1 when this group is named as minor suffix). The prefix is carboxy, the common name uses '-io', the IUPAC adds '–oic acid' and when directly connected to the ring system it is called 'carboxylic acid'. The carboxylic acid may be classified as 1°, 2° and 3° just like the alcohols, but generally it is not done frequently.

[a] Different types of acids:
[i] Aliphatic = R—COOH, [ii] Aromatic = Ar—COOH, [iii] Alicyclic—COOH.

Examples: Aliphatic carboxylic acids; common name and (IUPAC name)

[i] H—COOH = Formic acid (methanoic acid, ∵ C_1 acid, cf methane, C_1)

[ii] Cl—COOH = Chloroformic acid (Chloromethanoic acid)

[iii] CH_3—CO_2H = Acetic acid, (ethanoic acid, ∵ C_2 acid, cf ethane, C_2)

[iv] Cl_3C—CO_2H = Trichloroacetic acid (trichloroethanoic acid), no locant.

[v] HO_2C—CO_2H = Oxalic acid, (ethanedioic acid, 'e' stays)

[vi] $CH_3(CH_2)_3$—CO_2H = Valeric acid, (pentanoic acid).

[vii] 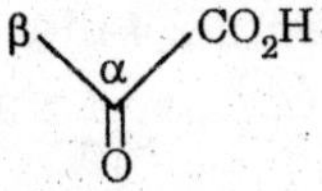

= α-Ketopropionic acid (2-oxopropanoic acid) (Pyruvic acid)

(Geneva nomenclature; note 'ionic acid')

[viii] 

= (3-oxopropanoic acid) (note 'anoic' acid).

[ix] 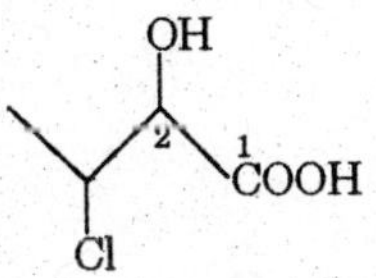

= α-Hydroxy-β-chlorobutyric acid (Geneva name) (α > β)

(3-Chloro-2-hydroxybutanoic acid) (alphabetic order)

[x] (a) 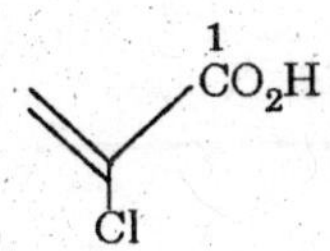

= (2-Chloropropenoic acid)

(b) 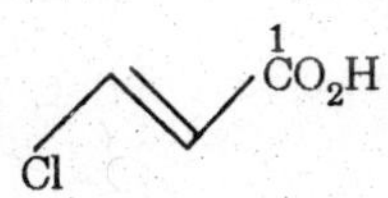

= (3-Chloropropenoic acid)

(c) 

= (Propynoic acid)

(d) 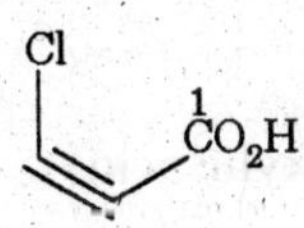

= (Chloropropynoic acid) (no locant)

[xi] 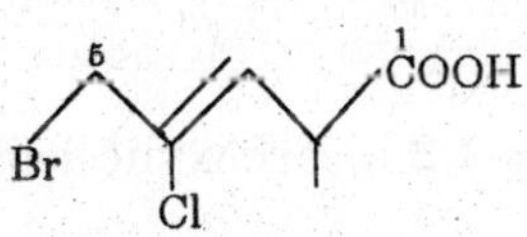

= (5-Bromo-4-chloro-2-methylpent-3-enoic acid)

[xii] 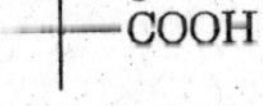

= Pivalic acid (Dimethylpropanoic acid) (no locant)

[xiii] 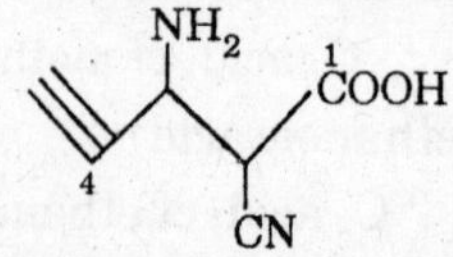 = 3-Amino-2-cyanopentynoic acid

('a' in penta is cancelled because 'y' follows)

Note: 'yne' needs no locant as this is the only position where it can be fixed.

[xiv]  = 2,4-Pentadienoic acid or penta-2,4-dienoic acid

[xv]  = (Pent-4-enoic acid), 'a' in penta is dropped.

[xvi]  = (Pent-2-enynoic acid);

Dropped 'a'—'e'—'e'.

 = (Pent-4-enynoic acid)

[b] Alicyclic acids : The COOH group does not retain the C_1 locant, but the connecting point of the cycle is marked as 1.

[i] 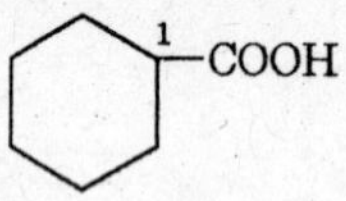 = Cyclohexanecarboxylic acid

[ii] 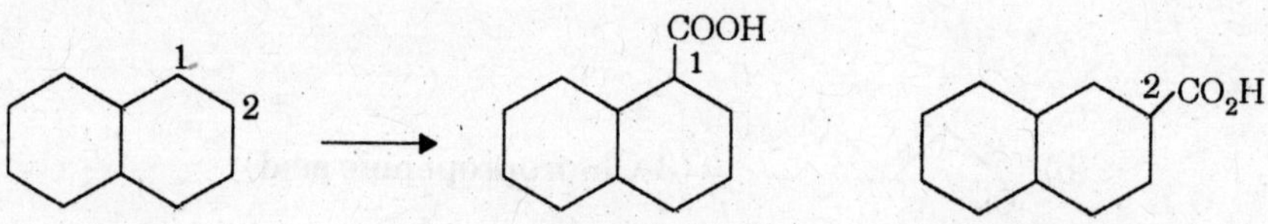

decalin — 1-decalincarboxylic acid — 2-decalincarboxylic acid

[c] Aromatic acids : The COOH group does not retain the C_1 locant.

[i] 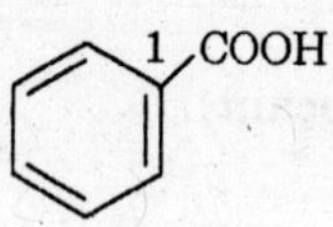 = Benzoic acid (benzenecarboxylic acid)

[ii] 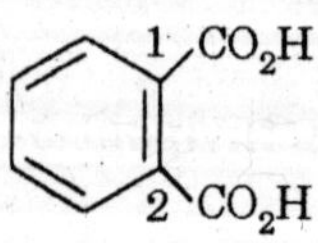 = Phthalic acid (benzene-1,2-dicarboxylic acid)

[iii] 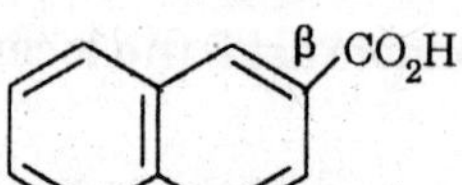 = β-Naphthoic acid (naphthalene-2-carboxylic acid)

[d] Arylalkanoic acids

[i] 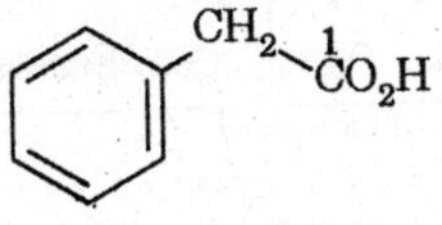 = Phenylacetic acid (phenylethanoic acid)

[ii] Ph—$\overset{\beta}{C}H_2$—$\overset{\alpha}{C}H_2$—$\overset{1}{C}OOH$ = β-Phenylpropionic acid (3-phenylpropanoic acid)

[e] Naming of carboxylic acids with stereochemical notations

[i] 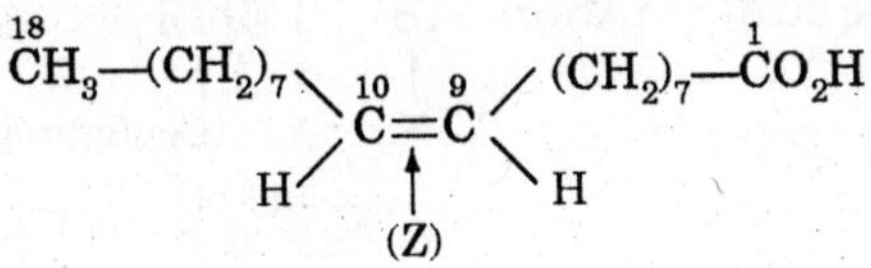 = (Z)-9-Octadecenoic acid (IUPAC)

or (Z)-Octadec-9-enoic acid (IUPAC).

[ii] 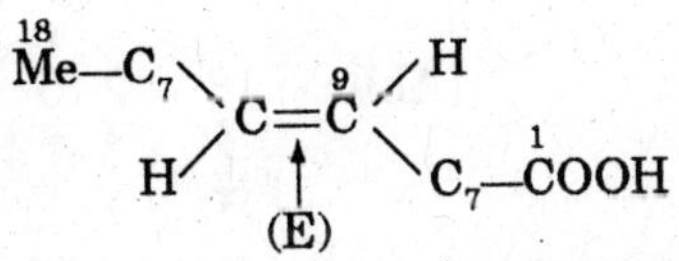 = (E)-9-Octadecenoic acid (IUPAC)

(Note: [i] and [ii] are diastereomers.)

[f] Abbreviated symbols of some olefinic acids : The symbol is x:y (z *c* or *t*)
where x = *total* number of carbons in the chain,
y = number of *olefinic* bonds (i.e., C=C) in the chain,
z = locant of C=C with the *cis/trans* configurational notation.

Examples

[A] Saturated Aliphatic Acid

Sl. No.	General formula	Common name	Abbreviation	IUPAC name
	$Me(CH_2)_n$—COOH	Aliphatic acid	(Total carbon)	Alkanoic acid
[i]	$n = 4$	Caproic acid	6 : 0	Hexanoic acid
[ii]	$n = 6$	Caprylic acid	8 : 0	Octanoic acid
[iii]	$n = 8$	Capric acid	10 : 0	Decanoic acid
[iv]	$n = 10$	Lauric acid	12 : 0	Dodecanoic acid
[v]	$n = 12$	Myristic acid	14 : 0	Tetradecanoic acid
[vi]	$n = 14$	Palmitic acid	16 : 0	Hexadecanoic acidc
[vii]	$n = 16$	Stearic acid	18 : 0	Octadecanoic acid
			[x : y]	

[B] Unsaturated aliphatic acids :

[i] CH_3—$(CH_2)_7$—CH=CH$(CH_2)_7$—$\overset{1}{C}OOH$ Oleic acid (common name)
↑ *cis* (also Z)

[(Z)-Octadec-9-enoic acid] [Abbr. 18 : 1 (9 *c*)]

(18 = x; 1 = y; 9 = locant of C=C is z; *c* = Configuration)

[Note that the z (in lower case) is the *location* of the C=C, not the configurational notations.]

[ii] Me—C_7—CH=CH—C_7—$\overset{1}{C}$OOH

(CH=CH: *trans* (also E))

[(E)-Octadec-9-enoic acid] [Abbr. 18 : 1 (9 *t*)]

(18 = x; 1 = y; 9 = locant of C=C is z; *t* = Configuration)

[iii] 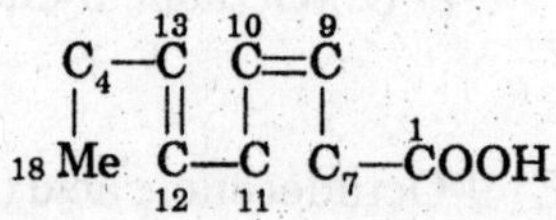

[Abbr. 18 : 2 (9 *c* . 12 *t*)]

(18 = x; 2 = y; 9 *c* = z, *cis* at C_9 (C=C); 12 *t* = z, *trans* at C_{12} (C=C))

[g] Nomenclature of polyhydroxy acids:

[i] Tartaric acid 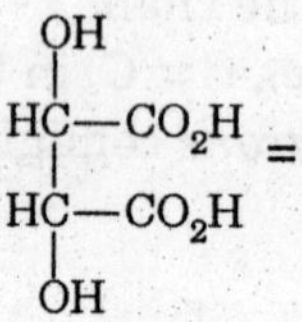 = (2,3-Dihydroxybutanedioic acid)

[ii] Citric acid 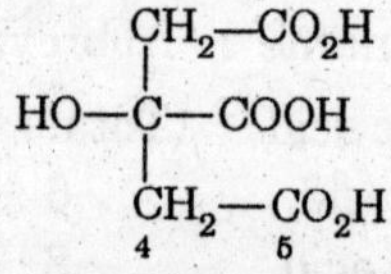 = 2-Hydroxypropane-1,2,3-tricarboxylic acid

= 3-carboxy-3-hydroxypentanedioic acid

[X] *Nomenclature of acid amides*: The functional group is $-CONH_2$, named carbamoyl as prefix and carbonamide as minor suffix, the carbon is always marked as 1 (IUPAC convention).

Examples

[i] 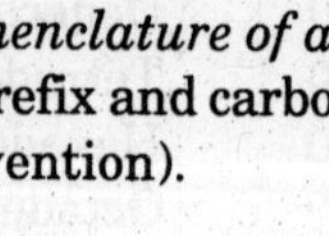 = Propanamide (IUPAC, 'oic acid' to 'amide')

[ii] 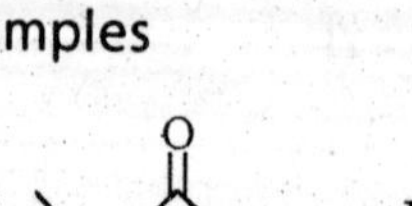 = 2-Methylbutanamide

(common 'ic' acid to amide as shown in, e.g., [x])

[iii] 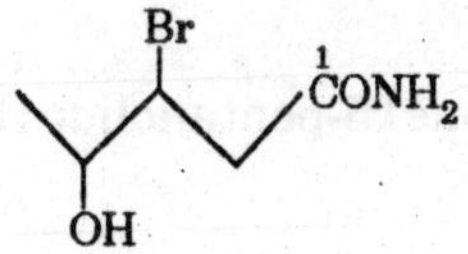 = **3-Bromo-4-hydroxypentanamide**

[iv] **H—C(=O)—NMe$_2$ = Dimethylformamide (DMF)**

[v] 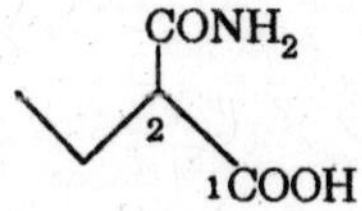 = **2-Carbamoylbutanoic acid (note $\overset{1}{C}OOH$ > $CONH_2$)**

[vi] **Ph—N(H)—C(=O)—N(H)—CH_3 = N-methyl-N[1]-phenylurea [urea is H_2N—C(=O)—NH_2]**

[vii] **H_2N—C(=O)—N(Ph)(Me) = N-methyl-N-phenylurea**

[viii] 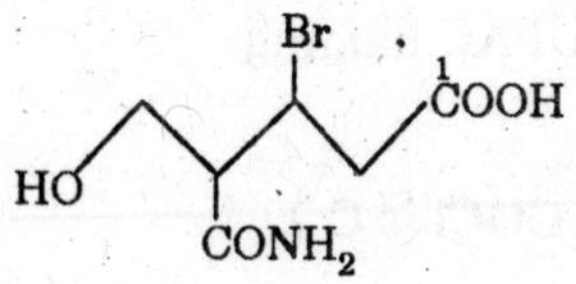 = **3-Bromo-4-carbamoyl-5-hydroxypentanoic acid**

B——c——h——C_5——$\overset{1}{C}OOH$

[ix] **H_2N—COOR = Alkyl carbamate (H_2N—COOH is carbamic acid)**
e.g., H_2N—$COOC_2H_5$ = Ethyl carbamate (ester of carbamic acid).

[x] **Acetamide (CH_3—$CONH_2$), benzamide (C_6H_5—$CONH_2$) are common names.**

[XI] ***Nomenclature of lactones*: They may be considered as internal esters, and called alkanolide (i.e., alkan~~e~~ +olide) (IUPAC).**

[i] 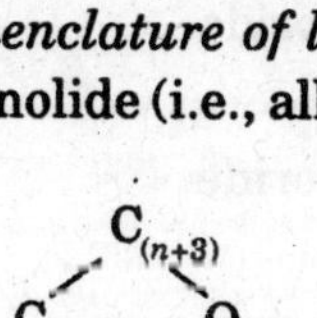 = **(n+3) Alkanolide**

when n = 1 → 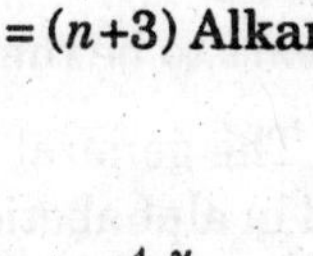 **= γ-Lactone (common name) 4-butanolide (IUPAC)**

$n = 2 \rightarrow$ 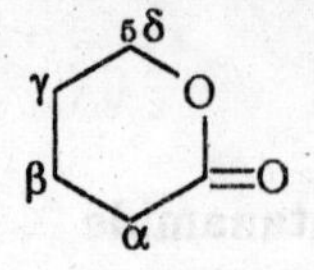 = δ-Lactone (common name) 5-pentanolide (IUPAC)

[ii] $n = 2 \rightarrow$ 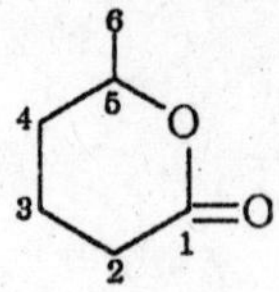 = 5-Hexanolide (derived from 5-hydroxyhexanoic acid)

[iii] 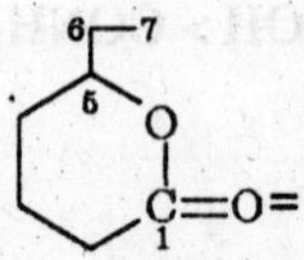 = 5-Heptanolide (derived from 5-hydroxyheptanoic acid)

[XII] *Nomenclature of acid halides*: The functional group is –COX (X-halogen), called carbonyl halide group. Thus, Cl—CO—Cl is called carbonyl chloride, the 'oic' acid is changed to 'oyl halide' in IUPAC naming and 'ic' acid is changed to 'yl halide' in the common names.

Examples

[i] CH_3—CO—Br = Acetyl bromide (common) and (ethanoyl bromide) in IUPAC naming.

[ii] 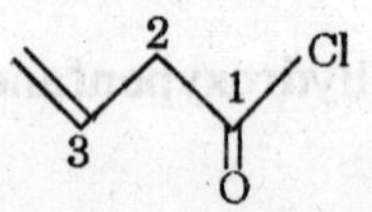 = But-3-enoyl chloride (the COCl is C_1).

[iii] CH_3—$(CH_2)_2$—COI = Butyryl iodide and (butanoyl iodide)

[iv] 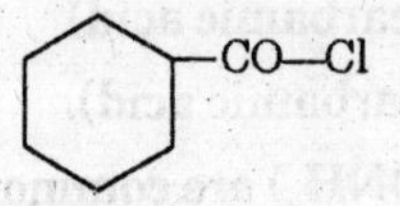 = Cyclohexanecarbonyl chloride.

[v] Br—CO—CH_2—COCl = Bromocarbonyl ethanoyl chloride

[vi] 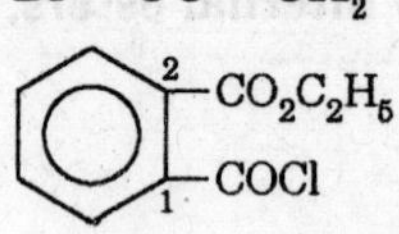 = 2-Carbethoxybenzenecarbonyl chloride

or 2-Chlorocarbonyl ethoxycarbonylbenzene
or Ethyl 2-chlorocarbonylbenzoate (appears to be the best nomenclature)

[XIII] *Nomenclature of carboxylic acid anhydrides*: The general formula is R—CO—O—CO—R′ (R=R′ or R≠R′, named in alphabetic order).

Examples

[i] CH_3—CO—O—CO—CH_3 (Acetic anhydride, common name) (Ethnoic anhydride, IUPAC name)

[ii] CH_3—CO—O—CO—CH_2—CH_3 (Acetic propionic anhydride, common name) (Ethanoic propanoic anhydride, IUPAC)

[iii] By 'oxo'/'oxa' nomenclature (IUPAC)

[a] $\overset{6}{C}H_3 - \overset{5}{C}O - \overset{4}{O} - \overset{3}{C}O - \overset{2}{C}H_2 - \overset{1}{C}OOH$

3,5-Dioxo-4-oxahexanoic acid *or* 3,5-Diketo-4-oxahexanoic acid

[b] $\overset{8}{C}H_3 - \overset{7}{C}O - \overset{6}{O} - \overset{5}{C}O\overset{4}{C}H_2\overset{3}{N}H - \overset{2}{C}H_2 - \overset{1}{C}OOH$

= 6-Oxa-5,7-dioxo-3-azaoctanoic acid (or 5, 7-Diketo-6-oxa-3-azaoctanoic acid)

[c] $HO_2C - \overset{2}{C}O - \overset{3}{O} - \overset{4}{C}O(CH_2)_3 - CH_3$ = 2,4-Diketo-3-oxaoctanoic acid ('ao' stays)

[d] $Et\overset{3}{C}O - \overset{4}{O} - \overset{5}{C}O - Pr^n$ = Butanoic propanoic anhydride (IUPAC) (3,5-Diketo-4-oxaoctane) 'ao' stays.

[XIV] *Nomenclature of esters*: The functional group is —COOR, called carbalkoxy or alkoxycarbonyl group when used as prefix. The 'ic' of the parent acid is changed into 'ate' in ester when used as minor suffix in both common and IUPAC system of nomenclatures. Thus,

[a] 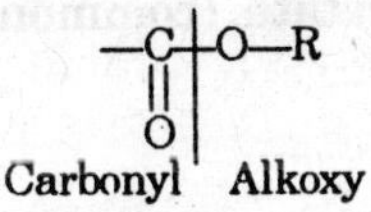

= carbalkoxy or alkoxycarbonyl

R—COOH → R—COO-R′
alkanoic acid → alkyl alkanoate
'ic' → (2 words) 'ate'

[b] When COOR is connected directly to a ring, then the naming is done as alkyl rootname carboxylate (not connected by hyphen).

Examples

[i] $H_3CCOOEt$ from Acetic acid (parent acid) (common name) → Ethyl acetate; (ester) (common name; 2 words)

from Ethanoic acid (parent acid) (IUPAC) → Ethyl ethanoate, (ester) (IUPAC) (2 words)

[ii] H—COOMe = Methyl formate → Methyl methanoate
(Methyl ester of formic acid) (common name) (IUPAC)

[iii] H—COOH

Formic acid or Methanoic acid $\xrightarrow[+\text{'ate'}]{-\text{'ic'}}$ $H—COOC_6H_5$ → Phenyl formate (common) or (Phenyl methanoate)

[iv] Pr^n—COOH

Butyric acid or Butanoic acid $\xrightarrow[+\text{'ate'}]{-\text{'ic'}}$ $Pr^n—COOPr^n$ → Propyl butyrate (common) or Propyl butanoate

[v] 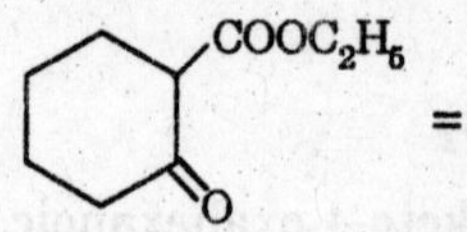 = Ethyl cyclohexanone-2-carboxylate

or 2-Carbethoxycyclohexanone

[vi] 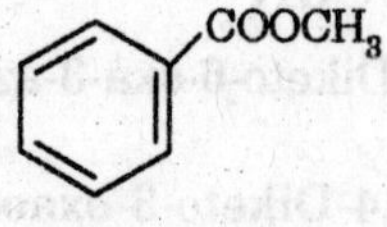 = Methyl benzoate

[vii] 

Salicylic acid → Methyl salicylate (common)

Methyl 2-hydroxybenzoate (IUPAC)

[viii] H—CO_2Et = Ethyl methanoate
H_3C—CO_2CH_3 = Methyl ethanoate
IUPAC

[ix] [a] Et—CO—OEt = Ethyl propanoate (IUPAC); Ethyl propionate (common)
[b] Me—CO—OPr = Propyl ethanoate; Propyl acetate. (common)

[x] $\overset{4}{C}H_3—\overset{3}{C}H(OH)—\overset{2}{C}H(Me)—\overset{1}{C}O_2Me$ =

(—CO_2R has the locant C_1)

Methyl ↑ 3-hydroxy-2-methylbutanoate
(No hyphen because Me is not connected directly to the carbon of the main chain)

[xi] 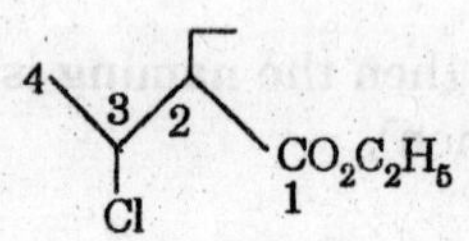 = Ethyl 3-chloro-2-ethylbutanoate

[xii] 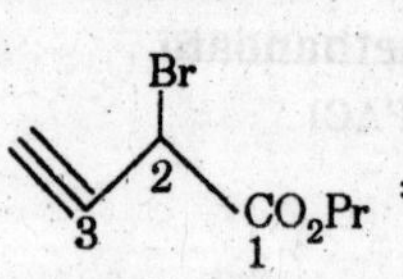 = Methyl 2-methylbut-3-enoate

('a' in buta and 'e' in ene are elided)

[xiii] 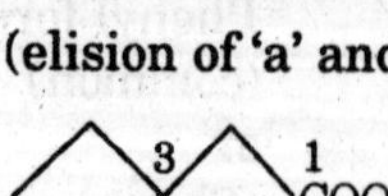 = Propyl 2-Bromobutynoate

(elision of 'a' and 'e', no locant for yne)

[xiv] 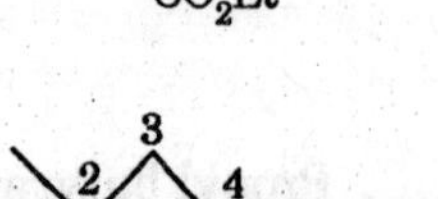 = 3-Carbethoxypentanoic acid (vide $\overset{1}{C}OOH$ > COOR)

[xv] 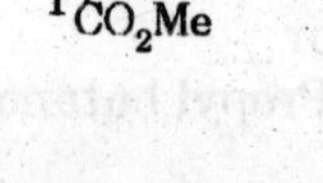 = Methyl 2-methyl-4-oxobutanoate (vide $\overset{1}{C}O_2R$ > CHO)

[xvi] 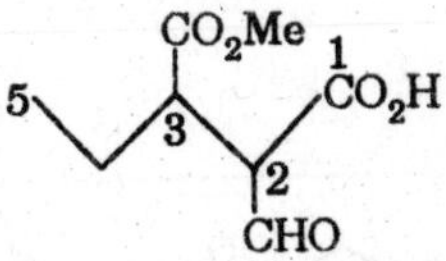 = 3-Carbomethoxy-2-formylpentanoic acid

or, 2-Formyl-3-methoxycarbonylpentanoic acid

$$\left(\underset{\text{Suf.}}{\overset{1}{C}O_2H} > \underset{\text{Pr.}}{CO_2Me} > \underset{\text{Pr.}}{CHO}\right)$$

64. Construction of the Structural Formula from IUPAC Name

[i] In the *first step* the root name of the given compound is determined, then it is drawn in short-hand formula.

[ii] In the *second step* carbons of the chain are marked by Arabic numerals.

[iii] In the *third step* the substituents are converted into respective formula from their names and attached at the proper location on the chain already marked by Arabic numerals.

[iv] In the *fourth step* the principal (i.e., main) suffix is fixed on the carbon chain at the specified location from the given nomenclature. These major suffixes are 'ene', 'yne' and 'enyne'; the 'ane' has no locant.

[v] The short-hand formula, thus obtained, should be named by the IUPAC naming method and checked with the given name to confirm the correctness of the derived formula. If this short-hand formula is found to be correct, then the long-hand formula can be drawn, if required. A little practice will enable the beginners to avoid all these stepwise procedure to write the formula directly from the given IUPAC nomenclatures.

[**Notes**: (a) The steps [iii] and [iv] may be reversed, if needed.

(b) The beginners should write down the IUPAC nomenclatures from all the structural formulas given under [63,B; I through XIV], and derive the structures from these names. In this way they will develop confidence and skill in naming the organic molecules varifying the correctness of the structure.

(c) They should memorize the common names of the low-molecular weight compounds (specially C_1 to C_6, as many as possible) (vide 67).

(d) They should be able to judge when the locants are unnecessary as the structures could be derived without them (e.g., methyl dimethylpropanoate, butenyne etc.)].

Examples

[i] 4-Carboxycyclohex-2-enone

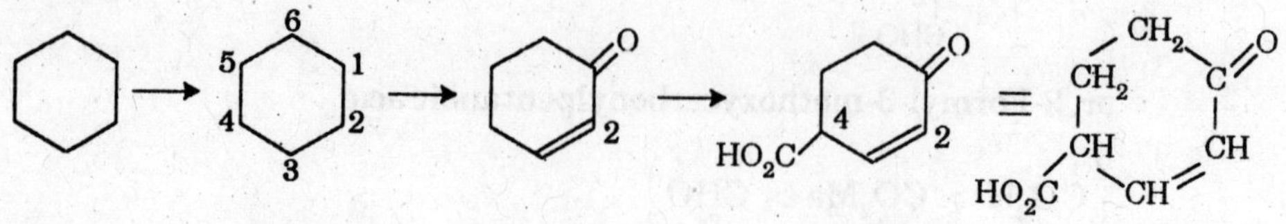

1st step (Cyclohexane root name) **2nd step (numbering)** **3rd step (Cyclohex-2-enone)** **4th step (short-hand)** **(long-hand)**

[ii] 4-Carbethoxy-4-methylcyclohex-2-enone

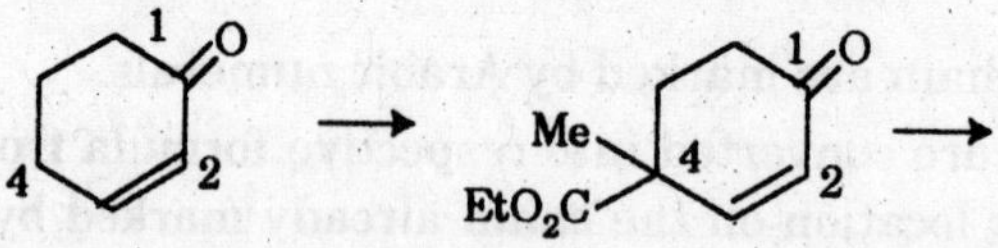

long hand, if required.

[iii] 3-Ethenylhepta-4,6-dienyne (in short-hand structure).
(The 'yne' needs no locant as C_1 is the only position left for triple bond).

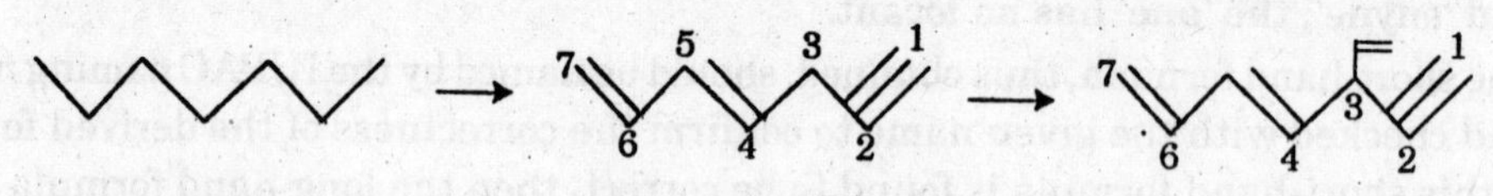

1st step (hepta, root name) **(Hepta-4,6-dien-1-yne) ('yne' needs locant)** **3-Ethenylhepta-4,6-dienyne (locant of 'yne' is no more needed)**

[iv] 4-Butyl-2,3-dimethyloctane (in short-hand structure)

(8-carbons) **(octane with locants)** **4-Butyl-2,3-dimethyloctane**

[v] 1-Bromo-5-hydroxyhept-6-yn-3-one

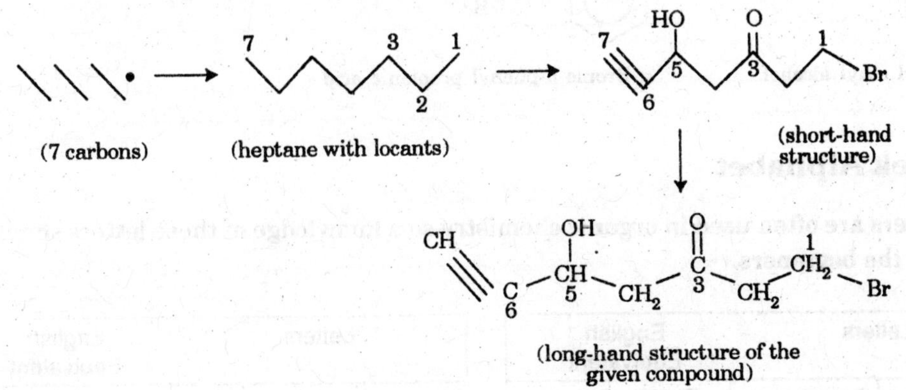

65. Derivatives of Benzene at a glance

The common names are given.

Benzene

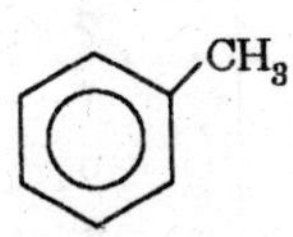

Toluene

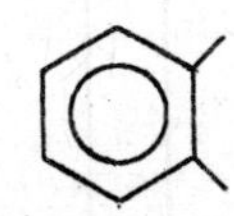

ortho-Xylene

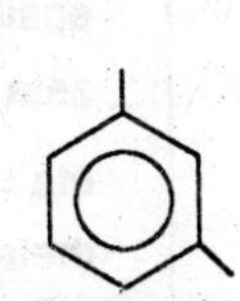

meta-Xylene

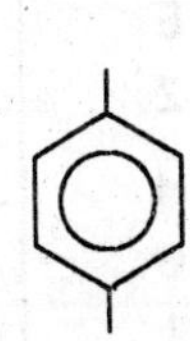

para-Xylene

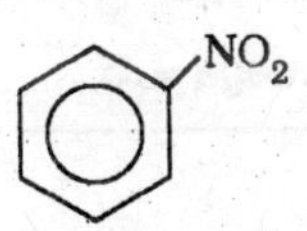

Nitrobenzene

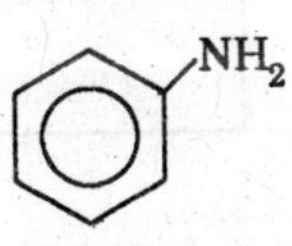

Aniline

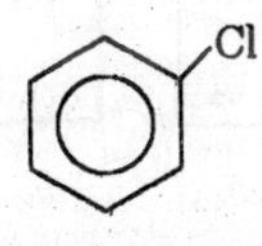

Chlorobenzene

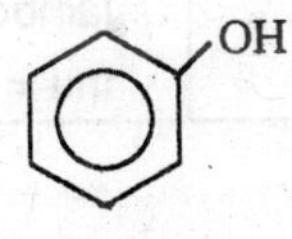

Phenol

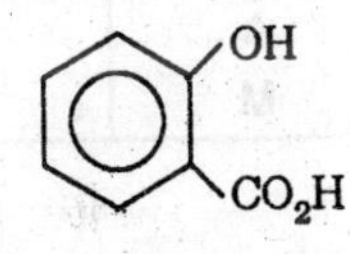

Salicylic acid

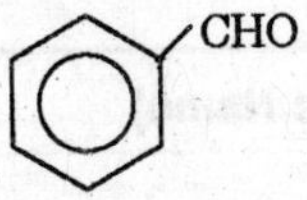

Benzaldehyde

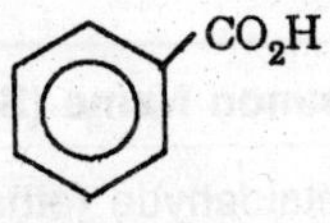

Benzoic acid

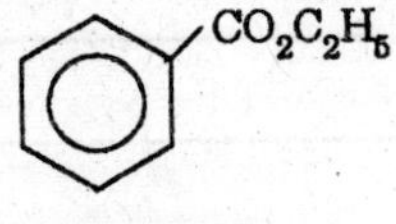

Ethyl benzoate

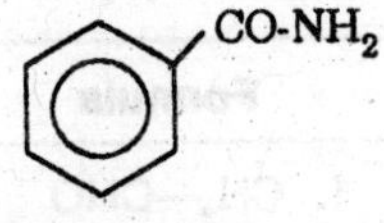

Benzamide

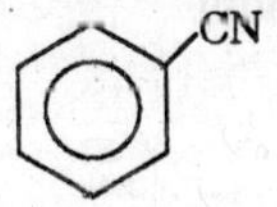

Benzonitrile

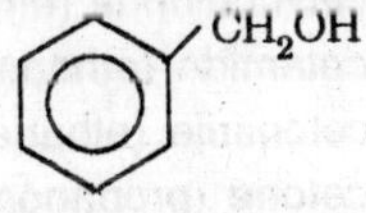

Benzyl alcohol

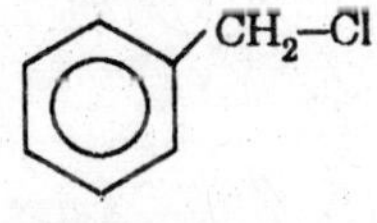

Benzyl Chloride

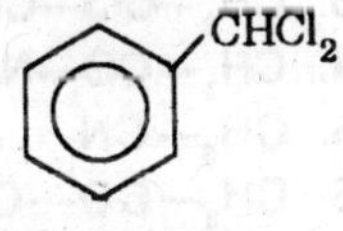

Benzal Chloride

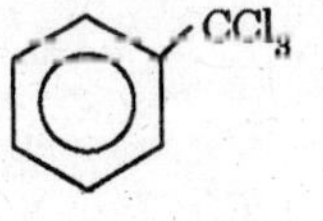

Benzotrichloride

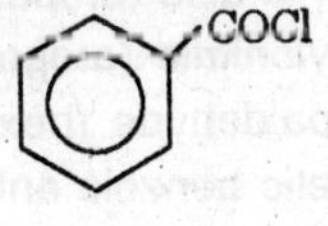

Benzoyl Chloride

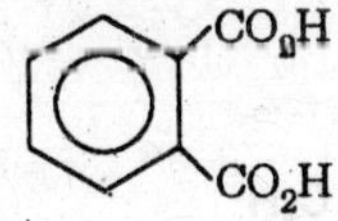

Phthalic acid

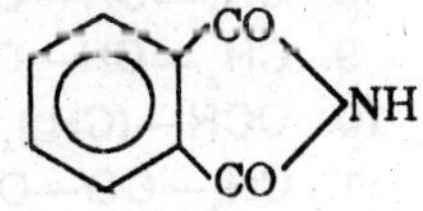

Phthalimide

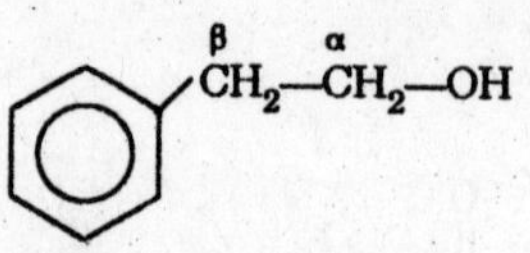

β-Phenyl ethyl alcohol

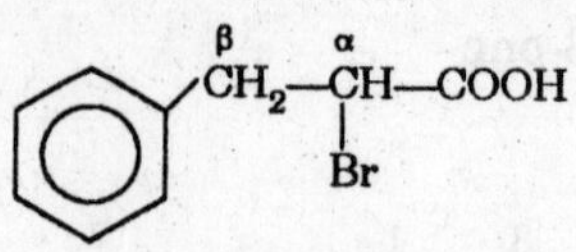

α-Bromo-β-phenyl propionic acid

66. Greek Alphabet

These letters are often used in organic chemistry so a knowledge of these letters should be helpful to the beginners.

Letters		English Equivalent
Higher case	Lower case	Pronunciation
Α	α	alpha = a
Β	β	beta = b
Γ	γ	gamma = g
Δ	δ	delta = d
Ε	ε	epsilon = ĕ
Ζ	ζ	zeta = z
Η	η	eta = ē
Θ	θ	theta = th
Ι	ι	iota = i
Κ	κ	kappa = k
Λ	λ	lambda = l
Μ	μ	mu = m

Letters		English Equivalent
Higher case	Lower case	Pronunciation
Ν	ν	nu = n
Ξ	ξ	xi = x
Ο	ο	omicron = o
Π	π	pi = p
Ρ	ρ	rho = r, rh
Σ	σ	sigma = s
Τ	τ	tau = t
Υ	υ	upsilon = u
Φ	φ	phi = ph
Χ	χ	chi = kh
Ψ	ψ	psi = ps
Ω	ω	omega = ō

67. The Common and IUPAC Names of Some Organic Compounds

Formula	Common Name (Systematic Name)
1. CH_3-CHO	1. Acetaldehyde (ethanal)
2. CH_3-COOH	2. Acetic acid (ethanoic acid)
3. $CH_3-CO-Cl$	3. Acetyl chloride (ethanoyl chloride)
4. $CH_3-CO-NH_2$	4. Acetamide (ethanamide)
5. CH_3-CN	5. Acetonitrile (ethanenitrile)
6. $CH_3-CO-CH_3$	6. Acetone (propanone)
7. $CH_2=CH-CHO$	7. Acrolein (propenal)
8. $CH_2=CH-COOH$	8. Acrylic acid (propenoic acid)
9. $CH_2=CH-CN$	9. Acrylonitrile (propenenitrile)
10. $OCH-(CH_2)_4-CHO$	10. Adipaldehyde (hexanedial)
11. $CH_3-CO-O-CO-C_6H_5$	11. (Acetic benzoic anhydride)
12. $CH_2=C=CH_2$	12. Allene (propadiene)
13. $CH_2=CH-CH_2OH$	13. Allyl alcohol (prop-2-en-1-ol)

Formula	Common Name (Systematic Name)
14. $CH_2{=}CH{-}CH_2SH$	14. Allyl mercaptan (prop-2-ene-1-thiol)
15. 	15. α-Amylene (pent-1-ene)
16. 	16. β-Amylene (pent-2-ene)
17. $C_6H_5NH_2$	17. Aniline (phenylamine)
18. $C_6H_5{-}O{-}CH_3$	18. Anisole (methoxybenzene)
19. $C_6H_5{-}CHO$	19. Benzaldehyde (benzenecarbaldehyde)
20. $C_6H_5{-}COOH$	20. Benzoic acid (benzenecarboxylic acid)
21. $C_6H_5{-}CO{-}NH_2$	21. Benzamide (benzenecarbonamide)
22. $C_6H_5{-}CN$	22. Benzonitrile (Benzenecarbonitrile, also phenyl cyanide)
23. $C_6H_5{-}CO{-}C_6H_5$	23. Benzophenone (diphenylmethanone)
24. $C_6H_5{-}CH_2{-}COOH$	24. Phenylacetic Acid (Phenylethanoic Acid)
25. $C_6H_5{-}PO_3H_2$	25. Benzene phosphonic aoid
26. $C_6H_5{-}SO_3H$	26. Benzene sulphonic acid
27. $C_6H_5{-}CO{-}Cl$	27. Benzoyl chloride
28. 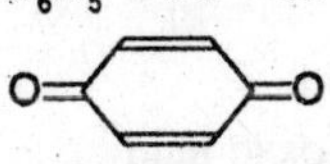	28. p-Benzoquinone (1,4-dihydro-1,4-benzenedione)
29. $C_6H_5{-}CH_2{-}OH$	29. Benzyl alcohol (phenylmethanol)
30. 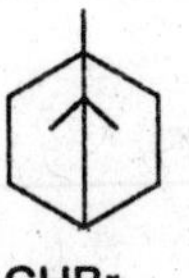	30. Bornane (bicyclo [2:2:1] 1,7,7-trimethylheptane)
31. $CHBr_3$	31. Bromoform (tribromomethane)
32. $C_6H_5{-}C_6H_5$	32. Biphenyl
33. $H_2N{-}(CH_2)_5{-}NH_2$	33. Cadaverine (1,5-diaminopentane)
34. 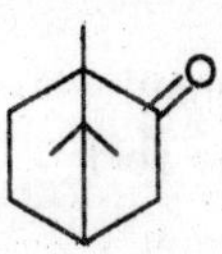	34. Camphor (bicyclo [2,2,1] 1,7,7-trimethylheptan-2-one)
35. $CH_3{-}(CH_2)_8{-}COOH$	35. Capric acid (decanoic acid)
36. $CH_3(CH_2)_4{-}COOH$	36. Caproic acid (hexanoic acid)
37. $CH_3(CH_2)_6{-}COOH$	37. Caprylic acid (octanoic acid)
38. $O{=}C(OH)_2$	38. Carbonic acid (hydroxymethanoic acid)
39. $C_6H_5{-}CH{=}CH{-}CHO$	39. Cinnamaldehyde (3-phenylpropenal)
40. $CH_3{-}CH{=}CH{-}CO_2H$	40. Crotonic Acid (but-2-enoic acid)
41. $CH_3{-}CH{=}CH{-}CN$	41. Crotononitrile (but-2-enenitrile)
42. $C_6H_5{-}CHMe_2$	42. Cumene (2-phenylpropane)
43. *p*-$MeC_6H_4{-}CHMe_2$	43. *p*-Cymene (4-isopropyltoluene)
44. 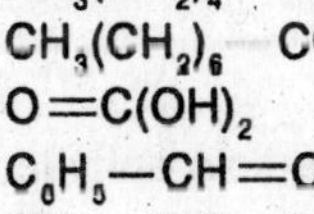	44. Cyclopentanecarboxylic acid
45. 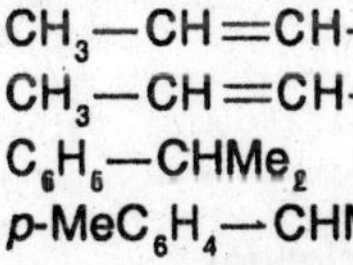	45. Cyclobutanecarboxamide

Formula	Common Name (Systematic Name)
46. $(p\text{-Cl-}C_6H_4)_2CH\text{—}CHCl_2$	46. DDD [1,1-dichloro-2,2-*bis* [(4′-chlorophenyl)-ethane.]
47. $(p\text{-Cl-}C_6H_4)_2\ CH\text{—}CCl_3$	47. DDT [1,1,1-trichloro-2,2-*bis* (4′-chlorophenyl)-ethane]
48. $C_6H_5NMe_2$	48. N,N-Dimethylaniline (N,N-dimethylphenylamine)
49. 	49. *gem*-dimethylcyclopropane (1,1-dimethylcyclopropane)
50. 	50. *vic*-dimethylcyclopropane (1,2-dimethylcyclopropane)
51. $CH_3\text{—}CH(OEt)_2$	51. Diethyl acetal (1,1-Diethoxyethane)
52. $C_2H_5\text{—}O\text{—}C_2H_5$	52. Diethyl ether (ethoxyethane or 3-oxapentane)
53. $CH_3\text{—}(CH_2)_5\text{—}COOH$	53. *n*-Enanthic acid (heptanoic acid)
54. $HO\text{—}CH_2\text{—}CH_2\text{—}OH$	54. Diethylene glycol (1,2-Ethanediol)
55. $H_2C{=}O$	55. Formaldehyde (methanal)
56. HCO_2H	56. Formic Acid (methanoic acid)
57. $HO_2C\text{—}CH{=}CH\text{—}CO_2H$ (*trans*)	57. Fumaric Acid (*trans*-butenedioic acid)
58. 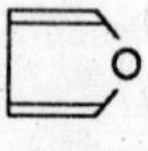	58. Furan (Oxole)
59. $HOCH_2\text{—}COOH$	59. Glycolic Acid (hydroxyethanoic acid)
60. $OHC\text{—}CO_2H$	60. Glyoxalic Acid (oxoethanoic acid)
61. $C_6H_{11}\text{—}COOH$ (Cyclo)	61. Hexahydrobenzoic Acid (cyclohexanecarboxylic acid)
62. $Me_3C\text{—}CH_2\text{—}CHMe_2$	62. *Iso*-octane (2,2,4-trimethylpentane)
63. $CH_2{=}C(Me)\text{—}CH{=}CH_2$	63. Isoprene (2-methylbutadiene)
64. $Me_2CH\text{—}CH_2\text{—}CHO$	64. *Iso*-valeraldehyde (3-methylbutanal)
65. $Me_2CH\text{—}CH_2\text{—}COOH$	65. *Iso*-valeric acid (3-methylbutanoic acid)
66. $CH_2{=}C{=}O$	66. Ketene (oxoethene)
67. $CH_3\text{—}CH(OH)\text{—}CO_2H$	67. Lactic Acid (2-hydroxypropanoic acid)
68. $HO_2C\text{—}CH{=}CH\text{—}CO_2H$ (*cis*)	68. Maleic Acid (*cis*-butenedioic acid)
69. $HO_2C\text{—}CH(OH)\text{—}CH_2\text{—}CO_2H$	69. Malic Acid (hydroxybutanedioic acid)
70. 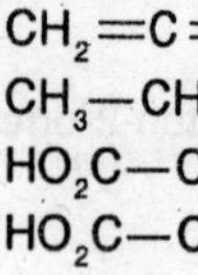	70. Maleic anhydride (*cis*-butenedioic anhydride)
71. 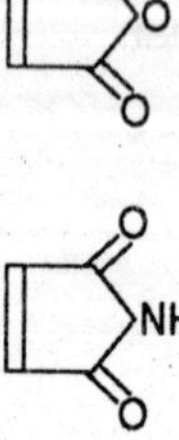	71. Maleimide (*cis*-butenedioicimide) (*pr.* cis.buten.dioic.imide)
72. $CH_2{=}CH\text{—}CH(OH)\text{—}CH_3$	72. Methyl vinyl carbinol (but-1-en-3-ol)
73. $HO_2C\text{—}CH_2\text{—}CO_2H$	73. Malonic acid (propanedioic acid)

Formula	Common Name (Systematic Name)
74. $MeO_2C—CH_2—CO_2Me$	74. Dimethyl malonate (dimethyl propanedioate)
75. 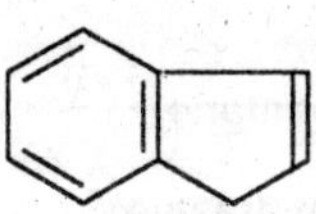	75. Indene (indene)
76. 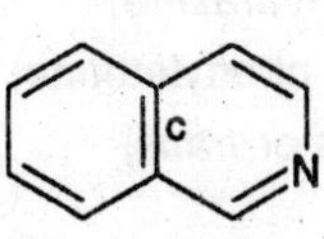	76. Isoquinoline (benzo[c]pyridine)
77. $Me_3C_6H_3$ (*sym*)	77. Mesitylene (1,3,5-trimethylbenzene)
78. $H_3C—C\equiv CH$	78. Methylacetylene (propyne)
79. 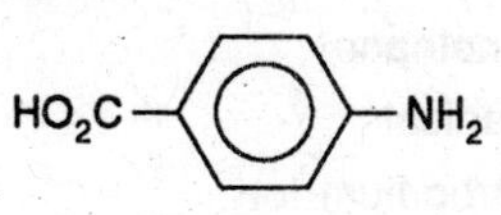	79. PABA = *p*-aminobenzoic Acid (4-aminobenzoic Acid)
80. 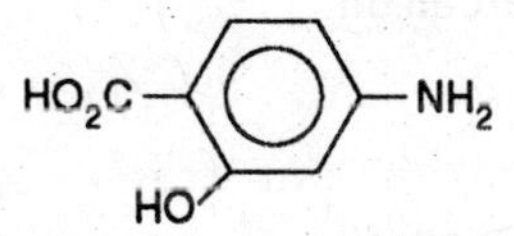	80. PAS = *p*-aminosalicylic Acid (4-amino-2-hydroxybenzoic Acid)
81. $C_6H_5O—C_2H_5$	81. Phenetole (ethoxybenzene)
82. C_6H_5OH	82. Carbolic Acid or Phenol (hydroxybenzene)
83. C_6H_4]—CO_2H, —CO_2H (*ortho*)	83. Phthalic Acid (1,2-benzenedicarboxylic acid)
84. 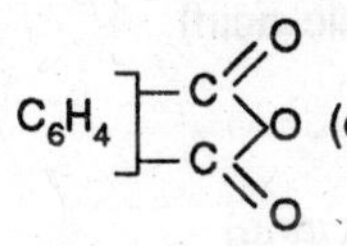 (ortho)	84. Phthalic Anhydride (1,2-benzenedicarbixylic anhydride)
85. 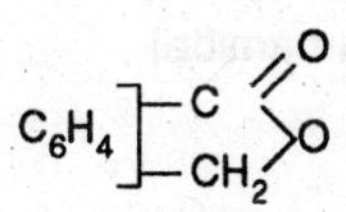	85. Phthalide (2,3-dihydro-2-oxa-1-indenone or 1(3) isobenzofuranone)
86. $HO—C_6H_2(NO_2)_3$ (2,4,6)	86. Picric Acid (2,4,6-trinitrophenol)
87. $Me_3C—CHO$	87. Pivalaldehyde (dimethylpropanal)
88. $Me_3C—CO_2H$	88. Pivalic Acid (dimethylpropanoic acid)
89. $Ph—CO—CH_2—CH_3$	89. Propiophenone (1-oxo-1-phenylpropane)
90. $H_2N—(CH_2)_4—NH_2$	90. Putrescine, also tetramethylenediamine (1,4-diaminobutane)
91. C_5H_5N	91. Pyridine (azine)
92. C_4H_4NH	92. Pyrrole (azole)
93. $CH_3—CO—CO_2H$	93. Pyruvic Acid (2-oxopropanoic acid)

Formula	Common Name (Systematic Name)
94. $C_6H_4\left]\begin{array}{l}-OH \\ -CO_2H\ (1, 2)\end{array}\right.$	94. Salicyclic Acid (2-hydroxybenzenecarboxylic-acid)
95. 	95. Spiropentane (spiro [2:2] pentane)
96. $H_2N-CONHNH_2$	96. Semicarbazide (carbamylhydrazine)
97. $C_6H_5-CH{=}CH_2$	97. Styrene (vinyl-benzene or phenylethene)
98. $MeCH{=}CH-Me$	98. *sym*-butene (but-2-ene; *cis* or *trans*)
99. 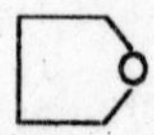	99. Tetrahydrofuran; THF (oxolane)
100. $C(NO_2)_4$	100. Tetranitromethane = TNM
101. Ph_3CH	101. Tritane (triphenylmethane)
102. Ph_3C°	102. Triphenylmethyl radical
103. $Ph_3C^{\oplus}$	103. Triphenylmethylcarbonium ion
104. $Ph_3C^{\ominus}$	104. Triphenylmethylcarbanion
105. 	105. Thiophen (thiole)
106. C_6H_5SH	106. Thiophenol (benzenethiol)
107. $CH_3-C_6H_2(NO_2)_3$ (*sym*)	107. Trinitrotoluene = TNT (2,4,6-trinitrotoluene)
108. $C_6H_3(NO_2)_3$ (*sym*)	108. Trinitrobenzene = TNB (1,3,5-trinitrobenzene)
109. Ph_3C-OH	109. Triphenylcarbinol (triphenylmethanol)
110. $C_6H_4\left]\begin{array}{l}-CH_3 \\ -CO_2H\end{array}\right.$ (*ortho*)	110. *ortho*-Toluic acid (2-methylbenzenecarboxylic acid)
111. 	111. Triethylenediamine (1,4-diazobicyclo [2,2,2] octane)
112. $O{=}C(NH_2)_2$	112. Urea (carbamide)
113. $MeNH-CONH_2$	113. N-methylurea (N-methylcarbamide)
114. $MeNH-CONHMe$	114. N,N′-Dimethylurea (N,N′-dimethylcarbamide)
115. $Me_2N-CO-NH_2$	115. N,N-Dimethylurea (N,N-dimethylcarbamide)
116. $Me(CH_2)_3-CO_2H$	116. Valeric Acid (pentanoic acid)
117. $H_2N-C_6H_3(CH_3)_2$ (2, 3)	117. 2,3-Xylide (2,3-dimethylphenylamine)
118. $CH_3-CHBr-CO_2H$	118. α-Bromopropionic Acid (2-bromopropanoic acid)
119. $(CH_3-CHBr-CO)_2O$	119. α,α′-Dibromopropionic acid anhydride (2,6-Dibromo-4-oxa-3,5-dioxoheptane)
120. $CH_3-CBr_2-CO_2H$	120. α,α-Dibromopropionic Acid (2,2-Dibromopropanoic Acid)
121. $BrCH_2-CHBr-CO_2H$	121. α,β-Dibromopropionic Acid (2,3-Dibromopropanoic Acid)

68. The Names and Formulas of a Few Groups (i.e., Radicals)

Formula	Common Name (Systematic Name)
1. $HC{\equiv}C-$	1. Acetenyl (ethynyl)
2. CH_3-CO-	2. Acetyl (ethanoyl)
3. $R-CO-$	3. Acyl (alkanoyl)
4. $CH_3-CO-O-$	4. Acetoxy (ethanoyloxy)
5. $CH_3-CO-CH_2-$	5. Acetonyl
6. $R-O-$	6. Alkoxy
7. $R-O-CO-$	7. Alkoxycarbonyl
8. $R-$	8. Alkyl
9. $CH_2{=}CH-CH_2-$	9. Allyl (prop-2-enyl)
10. $H_2N-\overset{\mid}{C}{<}^{C}_{C}$	10. Amino
11. $C_5H_{11}O-$	11. Amoxy (pentoxy)
12. $C_5H_{11}-$	12. *n*-Amyl (pentyl)
13. $C_nH_{2n-1}-$	13. (Alkenyl)
14. $C_nH_{2n-3}-$	14. (Alkynyl)
15. $Ar-CO-$	15. (Aroyl)
16. $Ar-$	16. (Aryl)
17. $ArO-$	17. (Aryloxy)
18. $>C{=}N-$	18. Anil
19. C_6H_5NHCO-	19. Anilide
20. $H_3C-O-C_6H_4-$	20. Anisal
21. $-N{=}N-$	21. Azo (diazo)
22. $C_6H_5-CO-O-$	22. Benzoyloxy
23. $C_6H_5CH{=}$	23. Benzal (benzylidene)
24. $C_6H_5C{\equiv}$	24. Benzo (benzylidyne)
25. C_6H_5CO-	25. Benzoyl
26. $C_6H_5-CH_2-$	26. Benzyl
27. $BrC_6H_4-SO_2$ (p)—	27. Bso $=$ Brosyl (4-bromobenzene sulphonyl)
28. $(CH_3)_3C-$	28. *tert.*Butyl
29. Me_3C-O-	29. *tert*-Butoxy
30. H_2N-CO-	30. Carbamoyl (amido)
31. $H_5C_2-O-CO-$	31. Carbethoxy
32. $>C{=}O$	32. Carbonyl=keto (oxo)
33. $CH_3-O-CO-$	33. Carbomethoxy *or* methoxycarbonyl
34. $HO-CO-$	34. Carboxy

Formula	Common Name (Systematic Name)
35. —O—CO—O—	35. Carbonyldioxy (dioxycarbonyl)
36. CH_3—CH=CH—CH_2—	36. Crotyl (2-butenyl)
37. $-N_2^{\oplus}X^{\ominus}$	37. Diazonium Salt
38. H—CO—O—	38. Formoxy
39. H—CO—	39. Formyl (methanoyl)
40. —C=C(OH)—	40. Enol
41. H_2N—N=C<	41. Hydrazo
42. HO—NH—CO—	42. Hydroxamic
43. HO—SO_3—O—	43. Hydrogen sulphate
44. —CO—NH—CO—	44. Imido
45. HN=C<	45. Imino
46. H_2N—SO_2—	46. Sulphonamide
47. Cl—SO_2—	47. Sulphonyl chloride
48. HO—N=C<	48. Oxime
49. —CH_2—	49. Methylene
50. >CH—	50. Methine
51. —CO—CH_2—CO—	51. Malonyl
52. H_3C—C<	52. C-methyl
53. H_3C—N<	53. N-methyl
54. C_6H_5—	54. Phenyl
55. C_6H_4=	55. Phenelene
56. H_3C—C≡C—	56. Propargyl (1-propynyl)
57. 	57. Isopentyl
58. 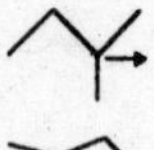	58. *Tert*-Pentyl
59. 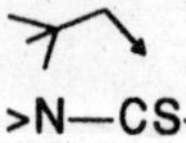	59. *Neo*-Pentyl
60. >N—CS—N<	60. Thiourea
61. MeC_6H_4—SO_2—(p)	61. Tso=Tosyl (*p*) (4-methylbenzenesulphonyl)
62. CH_2=CH—	62. Vinyl (ethenyl)
63. $CH_3C^{\oplus}$, ($R^{\oplus}$)	63. (Methylium) (alkylium)
64. $CH_3^{:\ominus}$, ($R^{\ominus}$)	64. (Methanide) (alkanide)

69. The Common and Systematic Names with Formula (Alphabetic Order) of Some Commercial Compounds

Commercial Name	Common Name (systematic name)	Formula
1. Acid of sugar	Oxalic Acid (ethanedioic acid)	$HO_2C—CO_2H$
2. Aqua fortis	Nitric Acid	HNO_3 $(HONO_2)$
3. Aqua rigia	Conc. HNO_3 + Conc. HCl (1:3 v/v)	$[HNO_3 + HCl]$
4. Aspirin	Acetylsalicylic Acid	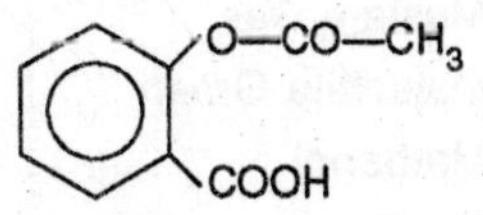
5. Bakelite	Phenol-formaldehye resin	C_6H_5OH/H_2CO polymer
6. Beet sugar	Sucrose	$C_{12}H_{22}O_{11}$
7. Benzine	Mixture of low boiling alkanes	———
8. Benzol	Benzene	C_6H_6
9. BHC; 666; Lindane	Gammexane (Hexachlorocyclohoxane)	$C_6H_6Cl_6$
10. Bone Black	Crude Animal Charcoal	C(crude)
11. Camphor	Camphor (Bicyclo [2, 2, 1] 1,7,7–trimethyl-2-oxoheptane)	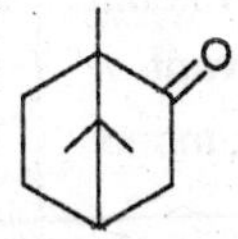 $= C_{10}H_{16}O$
12. Cane Sugar	Sucrose	$C_{12}H_{22}O_{11}$
13. Carbolic Acid	Phenol (hydroxybenzene)	C_6H_5OH
14. Chloramin T	Sodium *p*-toluenesulphochloramide	*p*-$H_3C—C_6H_4—SO_2\overset{\ominus}{N}ClNa^{\oplus}$
15. Corn Sugar	Glucose	$C_6H_{12}O_6$
16. Cream of tartar	Potassium Hydrogen Tartrate	$KHC_4H_4O_6$
17. CNG	Compressed Natural Gas	———
18. Cresylic Acid	Mixture of *o*, *m* and *p* cresols	$C_6H_4(OH)CH_3$ (*o*, *m* and *p*)
19. Dutch liquid	Ethylene Chloride (1,2-Dichloroethane)	$ClCH_2—CH_2—Cl$
20. Essence of mirbane	Nitrobenzene	$C_6H_5NO_2$
21. Formalin	40% aqueous solution of formaldehyde	Aq. H_2CO(40%)
22. Formin or Urotropin	Hexamethylenetetramin (Hexamine)	$(CH_2)_6N_4$
23. Freon or CFC	Chlorofluorocarbons	$(CFCl_3 + CHF_2Cl$ + etc)
24. Fruit sugar	Fructose	$C_6H_{12}O_6$
25. Fusel oil	Mixture of amyl alcohols	$C_5H_{11}OH$ (isomers)
26. Gasoline	Mixture of low boiling hydrocarbons suitable for internal combustion engines	———
27. Dextrose	Glucose	$C_6H_{12}O_6$
28. Glycerine	Glycerol (propane-1,2,3-triol)	$C_3H_5(OH)_3$ (1,2,3)

Commercial Name	Common Name (systematic name)	Formula
29. Grain alcohol	Ethyl alcohol (ethanol)	C_2H_5OH
30. Grape Sugar	Glucose	$C_6H_{12}O_6$
31. Lamp Black	Impure Carbon	C (Impure)
32. Lanolin	Mixture of cholesterol and esters.	———
33. Levulose	Fructose	$C_6H_{12}O_6$
34. Lysol	Cresol-soap solution	———
35. Mustard Gas	Mustard Gas, *bis*(2-chloroethyl) sulphide	$(Cl—CH_2—CH_2)_2S$
36. Malachite Green	A dye, colour is green	———
37. Methanol	Methyl alcohol (methanol)	CH_3OH
38. Metol	4-methylaminophenol sulphate	(p-$CH_3—NH—C_6H_4—OH$, $\frac{1}{2}$ H_2SO_4)
39. Metaldehyde	Polymer of acetaldehyde	$(CH_3CHO)_4$
40. Methylated spirit	Ethanol made unsuitable for drinking by adding methanol and pyridine (poison)	$(C_2H_5OH + CH_3OH + C_5H_5N)$
41. Milk sugar	Lactose	$C_{12}H_{22}O_{11}.\ H_2O$
42. Naphtha petroleum	Oil from distillation of petroleum	———
43. Naphtha solvent	Oil from distillation of coal-tar	———
44. Oil of bitter almond	Benzaldehyde	C_6H_5CHO
45. Oil of Garlic	Allylsulphide	$(CH_2=CH—CH_2)_2S$
46. Oil of Mirbane	Nitrobenzene	$C_6H_5NO_2$
47. Oil of mustard	Allyl isothiocyanate	$CH_2=CH—CH_2—NCS$
48. Oil of niobe	Methyl benzoate	$C_6H_5—CO_2CH_3$
49. Oil of pears	Amyl acetate	$CH_3—CO—O—C_5H_{11}$
50. Oil of pineapple	Ethyl butyrate	$CH_3—(CH_2)_2—COOC_2H_5$
51. Oil of wintergreen	Methyl salicylate	$C_6H_4(OH)—CO_2Me(o)$
52. Olefiant gas	Ethylene (Ethene)	C_2H_4
53. LPG	Liquified Petroleum Gas	———
54. Petroleum ether	Mixtures of Hydrocarbons, b.p. 40–60°C; and another at b.p. 60-80°C	———
55. Phenic acid	Phenol; carbolic acid	$C_6H_5—OH$
56. Paraldehyde	Paraldehyde	$(CH_3—CHO)_3$
57. Phosgene	Carbonyl chloride	$COCl_2$
58. Picric Acid	*sym.* Trinitrophenol (2,4,6-trinitrophenol)	C_6H_2 —OH (1), —$(NO_2)_3$ (2,4,6)
59. Power alcohol	Power alcohol	Mixture (EtOH +C_6H_6+Petrol)
60. Polyethylene	Polythene	$—(CH_2—CH_2)_n—$
61. Pyroligneous acid	Acetic acid (ethanoic acid)	$CH_3—CO_2H$

Commercial Name	Common Name (systematic name)	Formula
62. Pyroligneous spirit	Methyl alcohol (methanol)	CH_3OH
63. PVC	Polyvinyl chloride	Polymer
64. Quinol	Hydroquinone	*p*-HOC_6H_4—OH
65. Rectified spirit	Ethyl alcohol (90-95%)	$(EtOH + H_2O)$
66. Rochelle salt	Potassium sodium tartrate	$KNaC_4H_4O_6$, $4H_2O$
67. Saccharin	Saccharin (benzoic sulphimide)	C_6H_4 $\langle SO_2, CO \rangle$ NH (*ortho*)
68. Salol	Salol (phenyl salicylate)	C_6H_4 [—OH, —$CO_2C_6H_5$] (*ortho*)
69. Salt of amber	Succinic acid	HO_2C-$(CH_2)_2$—CO_2H
70. Salversion	3,3′-Diamino-4,4′-dihydroxy-arsenobenzene dihydrocloride	———
71. Spirit of wine	Ethyl alcohol	EtOH
72. Tartar emetic	Potassium antimonyl tartarate	$KSbC_4H_4O_7$, ½H_2O
73. Tetralin	Tetrahydronaphthalene	$C_{10}H_{12}$
74. Teflon	Teflon	—$(CF_2$—$CF_2)_n$—, Polymer
75. TNT	Trinitrotoluene	CH_3—$C_6H_2(NO_2)_3$ (2, 4, 6)
76. Toluol	Toluene	C_6H_5Me
77. Trioxan	Trioxan	$(CH_2O)_3$
78. Vanillin	Methyl ether of protocatualdehyde	C_6H_3 [—OH (1), —OCH_3 (2), —CHO (4)]
79. Wood alcohol or Wood naphtha	Methyl alcohol (methanol)	CH_3OH
80. Xylol	Xylene (*o*, or *m* or *p*)	C_6H_4 [—CH_3, —CH_3] (*o*, *m* and *p*)

70. Polymers

More than one small molecules combine to generate a very big molecule which has molecular weight as large as 2,50,000 or more. So it is almost impossible to write the exact molecular formula of such molecule called polymer. Thus when *n* number of molcculcs of ethylene ($CH_2{=}CH_2$) undergoes polymerisation, the product is called *polyethylene* (i.e., *polythene*) and shown as —$(CH_2$—$CH_2)_n$—. This type of polymerisation does not liberate smaller molecules like H_2O, NH_3, etc. One example is *n* number of molecules of *iso*-butylene polymerise to $(C_4H_8)_n$. When a polymer has general formula as $Z(X)_n$—Y when Z ≠ Y, and X is the repetitive unit, then such polymer is called telomer and the phenomenon is called *telomerism*. The ZY is present in the reaction mixture in smaller amount. The 'condensation polymerisation' differs from the former one in elimination of small molecules like H_2O, NH_3, etc. during condensation, one example is the condensation of glucose into starch.

The polymers are divided into three types, viz., (i) natural polymer e.g. rubber, protein, cellulose, starch, cotton, jute etc. These are bio-degradable hence they are not environment polluting agents, (ii) Semi-synthetic polymers like nitrocellulose, cellulose acetate etc. which involves the natural products as the starting materials and (iii) Synthetic or artificial polymers, like nylon , bakelite, etc. which cause environment polution because they are not biodegradable. Thermoplastics are linear polymers, soluble in organic solvents, hard when cold and soft when hot, and this property does not change. The other type is thermosetting, insoluble in organic solvents, hard on cooling but not soft on reheating but it undergoes thermolysis on strong heating. The examples are given in the following table.

Table: Commercial names and common names of a few plastics

[A] Thermosetting variety	Names: Class
[i] Araldite	[i] Epoxy resin
[ii] Bakelite	[ii] Phenol-formaldehyde
[iii] Catalin	[iii] Phenol-formaldehyde
[iv] Duralon	[iv] Furan
[v] Durite	[v] Phenol-furfural
[vi] Formica	[vi] Phenol-formaldehyde (laminated)
[vii] Indur	[vii] Phenol-formaldehyde
[viii] Melmac	[viii] Melamine-formaldehyde
[ix] Paraplex	[ix] Polyester
[x] Plaskon	[x] Melamine-formaldehyde, Polyester, Diallyl phthalate, Styrene modified, Urea-formaldehyde
[xi] Silastic	[xi] Silicone rubber
[xii] Textolite	[xii] Urea-formaldehyde
[xiii] Vibrin	[xiii] Polyester
[B] Thermoplastic variety	**Names: Class**
[i] Acrylon	[i] Ethyl acrylate
[ii] Bakelite	[ii] Polyethylene-polystyrene
[iii] Catalin	[iii] Polystyrene
[iv] Exon	[iv] Polyvinyl chloride, PVC
[v] Koroseal	[v] Modified chlorovinyl chloride
[vi] Nylon	[vi] Polyamide
[vii] Paradone	[vii] Coumarin-indene
[viii] Polythene	[viii] Polyethylene
[ix] Saran	[ix] Vinylidene chloride
[x] Styron	[x] Polystyrene
[xi] Teflon	[xi] Tetrafluoroethylene
[xii] Tenite	[xii] Polyethylene
[xiii] Valite	[xiii] Sugar cane resin
[xiv] Vuepak	[xiv] Cellulose acetate

[C] Gutta-percha and rubber are isomeric, $(C_5H_8)_n$, but the former is *trans* and the latter is *cis*. A little sulphur is added to the rubber to make it soft and durable. This is known as 'Vulcanisation' of rubber.

Examples

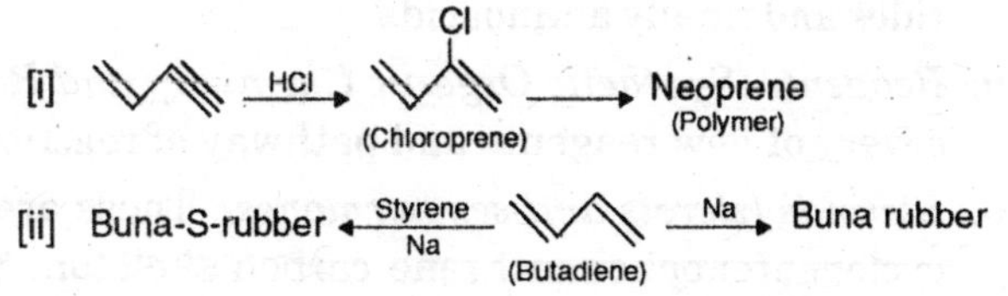

[iii] Isoprene $\xrightarrow[\text{Ziegler-Natta Catalyst}]{AlEt_3 + TiCl_3 + Li}$ Identical with natural rubber

71. Several Branches of Organic Chemistry

The following branches are noteworthy.

1. *Alkaloids*: These are nitrogenous basic plant products with noticeable physiological activity (e.g., quinine, coniine, morphin, etc.) Purines have *characteristic* carbon-nitrogen skeleton thus making them different class of compounds (e.g., caffein, uric acid).
2. *Anthocyanines*: These are natural glycosidic pigments found in coloured flowers (red, violet, yellow, etc). The sugar-free aglycons are termed anthocyanidins. These classes of compounds, are collectively called flavonoids (e.g., cyanidine, present in red roses, dahlias, etc.)
3. *Carbohydrates*: The general formula of this class of compounds is $C_nH_{2m}O_m$ ($n \geq m$). The lower molecular weight members are water soluble sweet crystalline polyhydroxy oxo compounds possessing chiral centres (e.g., glucose, $C_6H_{12}O_6$; sucrose, i.e., cane sugar, $C_{12}H_{22}O_{11}$, etc.). Cellulose and starch belong to this class of compounds when n is too large.
4. *Chemotherapy (and Drugs)*: The disease control by chemicals is chemotherapy and the agents are called chemotherapeutic drugs which includes antiseptic (e.g., dettol, iodoform, etc.), disinfectants (e.g., phenyl, cresol, etc.), analgesics (i.e., pain reliving, e.g., aspirin, brufen, etc.), antacids (e.g., omeprazole, etc.), antimicrobial drugs (sulphanilamide, sulphaguanidine, etc.), antibiotics (e.g., penicillin, streptomycin, etc.), antimaterials (e.g., paludrine, quinine, etc.), antipyretic (e.g., paracetamol, etc.) and tranquiliser.
5. *Dyes (and Indicators)*: Chemical used to impart colour to textile fabric like cotton, wool, silk, etc. (e.g., indigo, malachite green, etc.). *Indicators* changes colour with the change of pH of the medium mainly used in acid-base titration (e.g., methyl orange, congo red, etc.). Food colour is another aspect of the use of dyes (e.g., indigo-carmine, amaranth etc.). However, food preservaties and sweetening agents like saccharin do not come under this category.
6. *Instrumentations (and Analysis)*: Elemental analysis and applications of UV, IR, ^{1}HNMR, ^{13}CNMR, M.S., etc. are very interesting part of organic chemistry.
7. *Polymers (and Plastics, Rubber)*: This branch is very wide including natural polymers like cotton, jute, starch, etc. and synthetic polymers like nylon, terelene, etc. Rubber is a polymer derived from polymerisation of isoprene.
8. *Proteins (Polypeptides, Peptides and Amino-acids)*: These occur in the protoplasm of animal and plant cells and contain C, H, N and O, sometimes S and P (e.g., nucleo-

sides and Fe in haemoglobin). Proteins break down stepwise into polypeptides, peptides and finally aminoacids.

9. ***Reagents (Synthetic Organic Chemistry and Reaction Mechanism)*: This involves discovery of new reagents and pathway of reactions.**
10. ***Steroids (sterols and sex hormones)*: These are plant or animal products having 1,2-cyclopentenophenanthrene carbon skeleton. Steroids are crystalline hydroxy compounds or esters of fatty acids. Cholesterol is a zoosterol from animal, stigmasterol is a phytosterol from plants, and yeast, fungi, etc. delivers mycosterols. Sex hormones (e.g., testosterone, oestrone etc.) are members of this class also.**
11. ***Terpenoids (essential oils)*: These are plant products having general formula $(C_5H_8)_n$. The lower members are sweet smelling (jasmine oil, camphor etc.).**
12. ***Vitamins (and Hormones)*: These are minute quantity of chemicals essential to keep good health [e.g., vit A, B (complex), C, D, E and K]. Vitamins are not synthesised in body but hormones e.g., thyroxin, testosterone, etc.) are prepared in the body exerting great influence.**

72. Exercises

1. **Determine the number of carbon atoms marking each one of them as 1° (primary, *p*), 2° (secondary, *sec*), 3° (tertiary, *tert*) and 4° (quarternery, *neo*) in the following drawings, and convert each one into 'Long-hand' from the 'Short-hand' picture.**

[a] 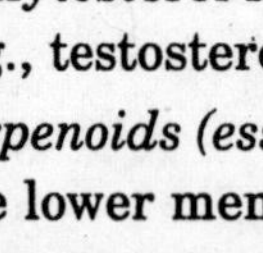

[b] 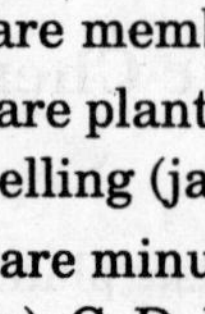

[c] 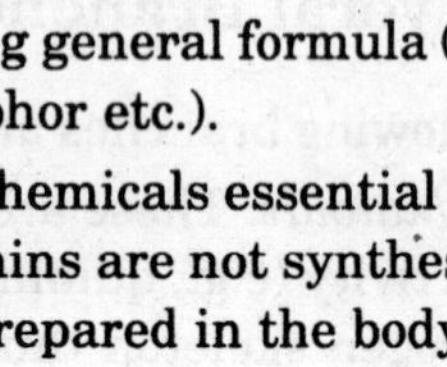

[d] 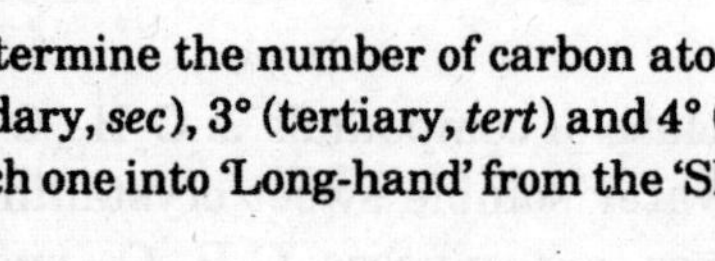

2. **Convert the following formula into 'Short-hand' formula.**

[a] 

[b] 

[c] CH_3—CHO

[d] $(CH_3)_3$C—OH

[e]
CHO
|
NC—CH_2—CH—CH_2—NC

3. **Draw the short-hand structure from the following condensed formulae:**

[a] $(CH_2—CO_2H)_2$, succinic acid;

[b] $(COOH)_2$, oxalic acid;

[c] CH_3—$(CH_2)_2$—$CBr(NO_2)$—CMe_3,

[d] Me—N(—Me)—CH_2—CH(OH)—CCl_3

4. Draw the Lewis-dot picture of the following:
[a] $MeCH_2—OH$, [b] $Me—CH=CH—NH_2$ [c] $HO—CH_2—CHCl—CH_2SH$
[d] $H_2N—CH_2—CH(OH)—CHO$ [e] $H_3C—C\equiv C—CH=CH—CN$
[f] $CH_3—O—CH_2—CH_2—Br$ [g] $(CO_2H)_2$ [h] $CN—CH_2—CN$

5. Does any one of the following skeletal formulae represent different compound?

[a] 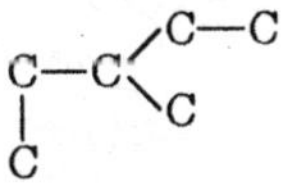 [b] 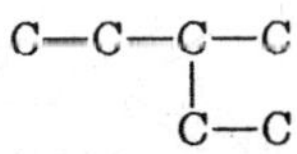 [c] 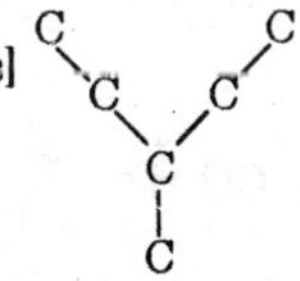

[d] 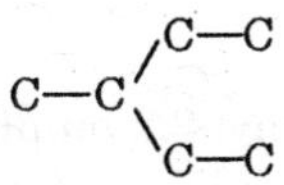 [e] 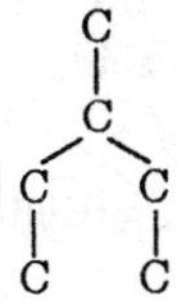

6. Draw *cis*- and *trans*-butene in longhand formula.

7. Draw stereostructural formula of C_3H_8.

8. In short-hand formula draw all the isomers of C_4H_{10}. Name each one of them.

9. Identify the *neo*-carbon from the following list of compounds:
[a] $CH_2=C=C(CH_3)_2$ [b] $CH_3—CCl(Br)—CO_2H$ [c] $HO_2C—CO_2H$
[d] $C(CH_2OH)_4$ [e] CCl_4

10. Identify the iso-compounds from the following list:

[a] $(CH_3—CH_2)_2CH(OH)$ [b]  [c] 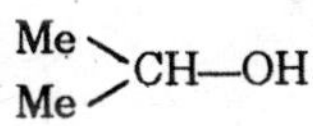

[d] 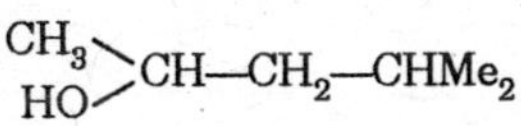 [e] 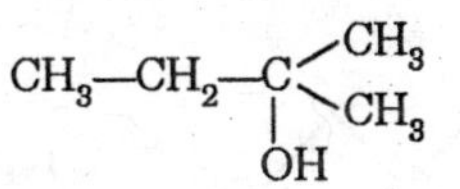 [f] 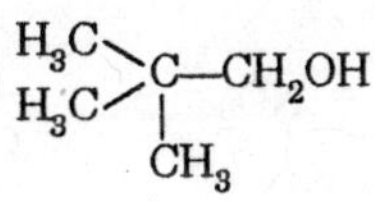

[g] 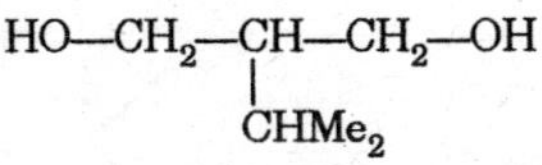

11. Give the IUPAC nomenclature of the following compounds:

[i]  [ii] 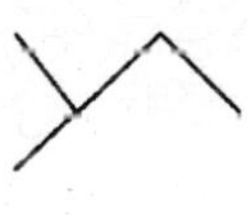 [iii] 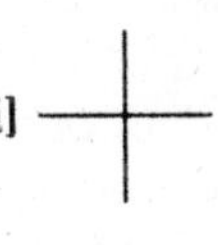

[iv] 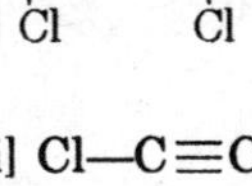 [v] [vi]

[vii] $Cl—C\equiv C—Cl$

12. Determine the prefix and suffix in the following pairs:
[i] Aldehyde and carboxylic acid, [ii] cyanide and alcohol, [iii] ene and yne, [iv] yne and one, [v] alkoxycarbonyl and sulphonic acid, [vi] carboxylic acid and carbethoxy group, [vii] ol and al, [viii] ol and amine.

13. Interchange the prefix and suffix in the following:
[i] Aminopropane, [ii] 2-Methyloxopropanenitrile.

14. Show the Roo-Su-Pre and locants of the following compounds:

[i] 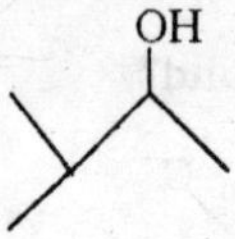

[ii] 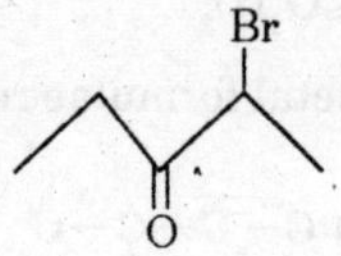

[iii] HOOC—COOH

[iv] $OCH—CO_2H$

[v] $OCH—C(=O)—CO_2R$

15. Draw the structure of the molecules from the following common names:
[i] α-Bromopropionic acid, [ii] β-Bromobutyric acid.

16. What is wrong in the following IUPAC names?
[i] Butanene, [ii] 2,2-Methylbutane, [iii] 22-Dimethylbutane, [iv] Dimethyl-2,2-butane, [v] Butane-2,2-dimethyl and [vi] Butol.

17. Give the short-hand structures of the following compounds:
[i] Propane, [ii] Propene, [iii] Propyne, [iv] Propanal, [v] Propanoic acid, [vi] Propenoic acid, [vii] Propynoic acid, [viii] Propanone and [x] Propan-2-ol.

18. Give the IUPAC and common names of the following compounds:

[i] 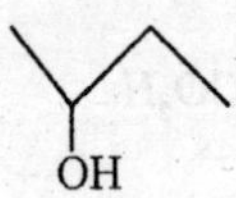

[ii] 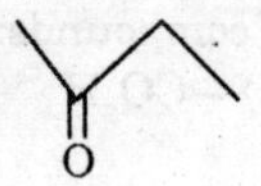

[iii] 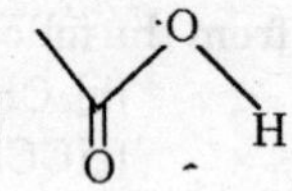

[iv] 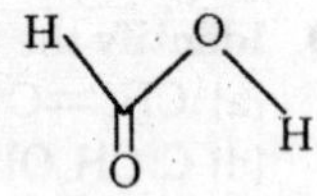

[v] 

[vi] 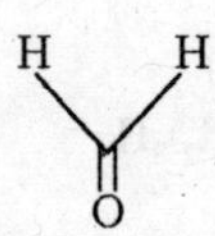

[vii] 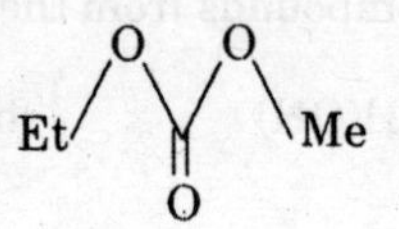

[viii] 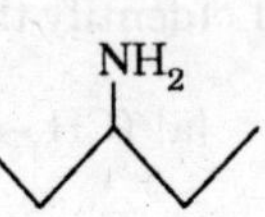

[ix] 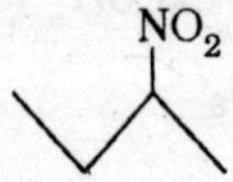

[x] 

[xi] 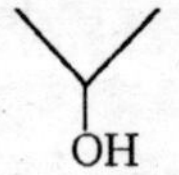

[xii] 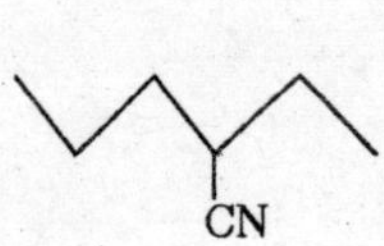

[xiii] 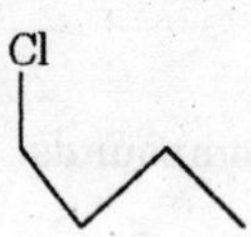

[xiv] 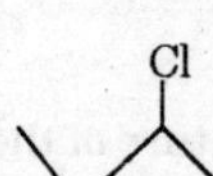

[xv] 

[xvi] 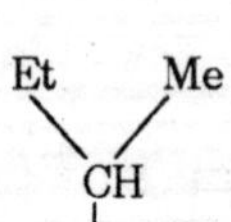

[xvii] 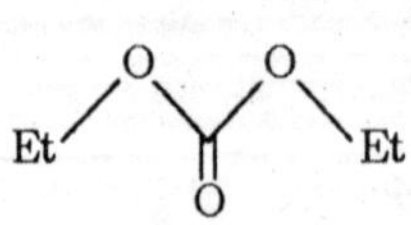

[xviii] 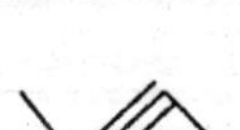

[xix] 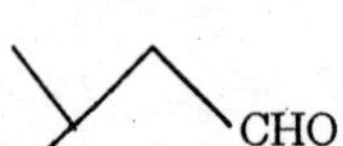

[xx] $HO_2C—CO_2H$

[xxi] $OCH—CO_2H$

[xxii] 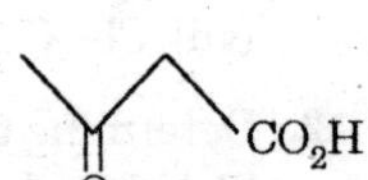

[xxiii] 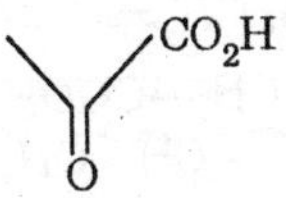 [xxiv] 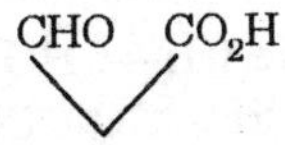 [xxv] 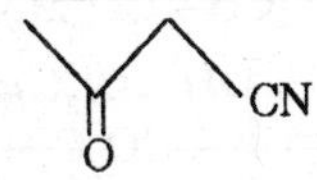 [xxvi] 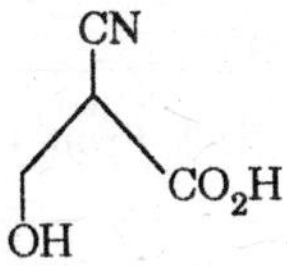

[xxvii] $CH_2(CO_2H)_2$ [xxviii] $(CH_2)_2(CO_2H)_2$ [xxix] $HO_2C—CH(OH)—CH(OH)—CO_2H$

[xxx] 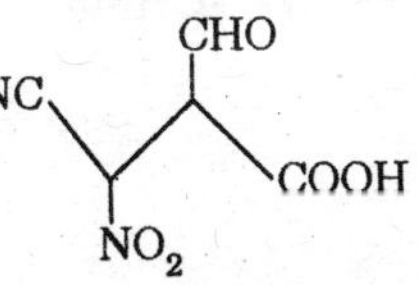 [xxxi] 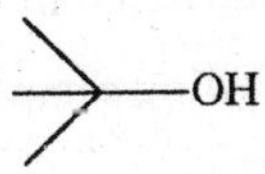 [xxxii] 

19. **Draw the structural formula of the following compounds:**
[i] 2-Aminobutane, [ii] 1-Butene, [iii] 1-Butyne, [iv] 2-Oxopentane, [v] 3-Cyanohexane, [vi] 3-Ketoheptane, [vii] Butanone, [viii] Ethyl 2-methylbutanoate, [ix] Ethyl propanoate, [x] Propyl ethanoate.

20. **What are wrong in the following IUPAC nomenclatures?**
[i] 2-Methylbutane, [ii] 2 3-dimethylbutane, [iii] 2,3 dimethylbutane, [iv] 2,3-di-methyl-butane, [v] propanene, [vi] 2-butaol, [vii] dichlorodiane, [viii] tetra-chlorotetrane, [ix] 1,1,2,2,2-pentachloropentane, [x] 2-chloro-3-bromobutane, [xi] chlorobutane, [xii] 1-chloroethane.

21. **Identify the compounds which may require the locants in the following:**
[i] Dichloroethane, [ii]dichloromethane, [iii] ethanol, [iv] trichloroethane, [v] butyne, [vi] propene, [vii] butanone, [viii] methylbutane, [ix] dimethylbutane, [x] butanoic acid, [xi] butenol, [xii] butenoic acid, [xiii] oxabutane, [xiv] butadiene, [xv] oxobutane, [xvi] cyanobutanoic acid, [xvii] tetramethylbutane.

22. **Draw the structure of the following compounds:**
[i] Methyl-1,3-butadiene, [ii] 2,3-dimethylbutane, [iii] neo-pentane, [iv] 4-methylcyclopentene, [v] 2-ethyl-3-methylpentanal, [vi] 2-methyl-2-butenal, [vii] 2-oxopropanoic acid, [viii] 4-oxopentanoic acid, [ix] ethanedioic acid, [x] 4,5-dimethylhex-2-en-1-ol, [xi] methoxypropane, [xii] 2,3,5-trimethyl-4-azahexane, [xiii] 2-bromo-3,3-dichlorobutanenitrile, [xiv] 3-amino-2-formylbutanoic acid, [xv] 2-hydroxy-3-methylbutanamide.

23. **Give the IUPAC nomenclature of the following compounds:**
[i] $Me_2CH—C{\equiv}CMe$, [ii] $Me_2C{=}CH_2$, [iii] $MeCCl_2—CH_2—CH_3$, [iv] $MeCO—CHMe_2$, [v] $MeCH{=}CH—CO_2H$, [vi] $OCH—COOH$, [vii] $EtCO—CH_2—CH(Me)—CHO$, [viii] $CH_3—CO—OEt$, [ix] $CH_3—CH_2—COOMe$, [x] $CH_3—CH(Cl)—CH(NO_2)—CMe_2—CH_2—COCH_3$, [xi] $CH_3—NH—CH_2—O—CH_2—CH(Me)—CHO$, [xii] $Et—C{\equiv}C—C({=}O)—Et$.

24. **[a] Give the common names and IUPAC nomenclatures of the following:**
[i] From $C_1 \rightarrow CHCl_3$, HCO_2H, H_2CO, $COCl_2$, HCO_2Me, HCBrClF, CBrCl FI

[ii] From $C_2 \rightarrow ClF_2C—CF_2Cl$, $Cl_2CH—CH_2Cl$, $CH_3—CHO$, OCH COOH, $HO_2C—CO_2H$, $Cl_3C—CO_2H$, $CH_3—CO_2C_2H_5$, 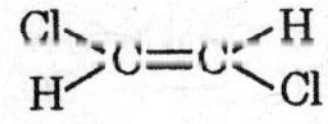**, $CH_2{=}CCl_2$, $Cl_3C—CHO$, $CH_2{=}CHOH$, $CH{\equiv}C—NO_2$, $BrC{\equiv}CCl$, $Cl_3C—CH_2—NO_2$.**

[iii] From $C_3 \rightarrow CH_3—CO—CO_2H$, $OCH—CH_2—CO_2H$, $ClCH_2—CH_2—CO_2C_2H_5$, $CH_3—CH(OH)CO_2H$, $CH_3—CHBr—CHO$, $O_2N—CH_2—CH(Cl)—CN$, Me_2CHOH,

OCH—CH_2—CH_2—OH, CH_2=CH—CHO, CH_3—CO—CH_2—OH, CH_2=CH—CO_2Me, CH≡C—CO_2Me, ClC≡C—CHO, CH_2=C=CH_2.

[iv] From C_4 → Me_3C—OH, $(Cl_3C)_3$C—OH, Me_2CH—CHO, CH_3CO—CH_2—COOH, CH_3—CH=CH—$CO_2C_2H_5$, OCH—CH_2—CO—CH_3, CH_2=C=CH—CH_3, CH_3—CH(Br)—CH(Cl)—CH_3, CH_3—CO—CH(Br)—CHO, CH≡C—CH=CH_2, CH≡C—CO—COOH, C_2H_5—O—C_2H_5, Me_2CH—O—CH_3, Me_2CH—NH—$CHMe_2$, Me_2N—CO—NHMe, MeCH_2—NH—CO—NH—$CHMe_2$.

[b] Identify the optically active compounds listed in [a].

25. Give the IUPAC nomenclature of the following compounds:

[i] 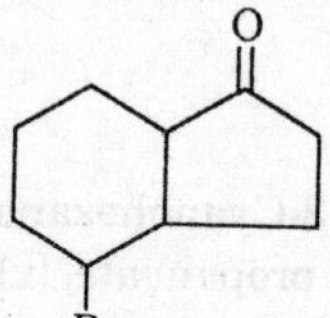

[ii] 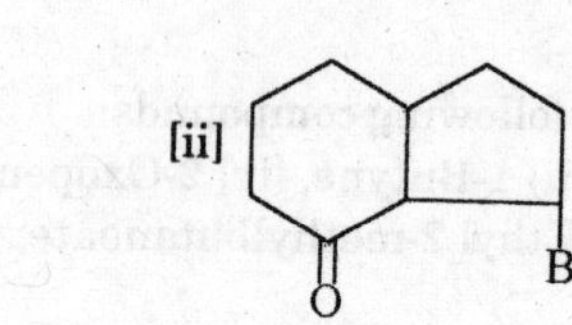

[iii] 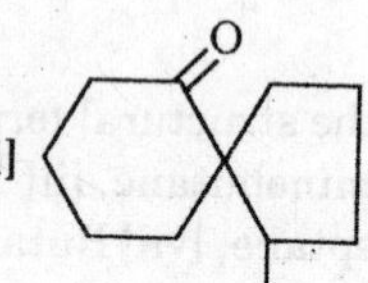

[iv] 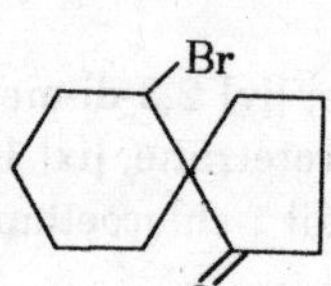

[v] —[3, 2, 0]—

26. [i] Does propanone require a locant?
[ii] Does propanol require a locant?

27. Transfer the prefix into suffix and vice versa of the following:
[i] 2-oxopropanoic acid, [ii] 2-chloroethanol.

28. What will be the names of prefix and suffix of each one of the following?
—CHO, —COOH, —$CO_2C_2H_5$, —CN, —CO_2Me, —$CONH_2$, —COCl, —OH, —NO_2

29. What is the significance of Fe(η^5—C_5H_5)?

30. Write the IUPAC name of HO_2C—$(CH_2)_3$—CH(COOH)—$(CH_2)_2$—COOH and

$HO_2C(CH_2)_2$—CH(CH_2—COOH)—CH_2—COOH (Note: —CH_2—CO_2H is carboxymethyl group).

31. Indicate the prefix(es) and suffix(es) in the following compounds:

[i] R [—NH_2, —OH]
[ii] R [—COOH, —CHO]
[iii] R [—Cl, —CN]
[iv] R [—Oxo, —Oxa]
[v] R [—CHO, —CN]
[vi] R [—NO_2, —OH]
[vii] R [—yne, —oxo]
[viii] R [—ene, —yne]
[ix] R [—Br, —I, —CHO]
[x] R [—COOH, —CO_2CH_3, —$CONH_2$]
[xi] R [—CHO, —COCl, —CO_2Me]
[xii] R [—Br, —Cl, —I]
[xiii] R [—NO_2, —ene, —OH]

32. What is wrong in the following names? Write the correct names.
[i] 2-carboxybutanal, [ii] methylmethanoate, [iii] 3-bromo-2-butanol, [iv] 2,2-dimethyl 3 butanol, [v] butane-2-ene, [vi] 4,4,4-trimethylbutanenitrile, [vii] 23-dibromo 6 hexanoic acid.

33. Consider the following compounds as the derivatives of hydrocarbons and write the systematic names. (The suffix will be either ane, or ene or yne or enyne.)

[i] 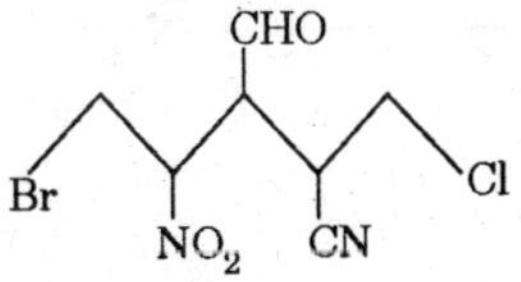

[ii] 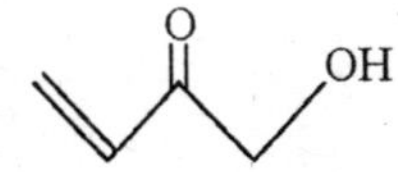

[iii] 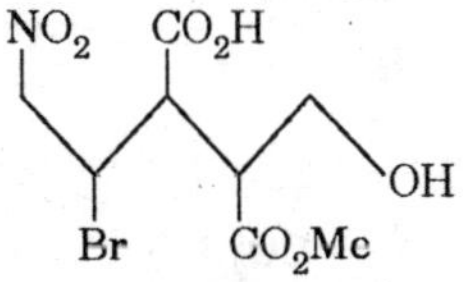

[iv] 
[v] 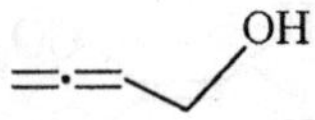

34. List the functional groups which must have the locant C_1. What is the minimum locant of a keto group in open chain compounds?

35. Arrange the following groups in the order of decreasing priority as suffix: —$CONH_2$, —Br, —CO_2R, —COCl, —COOH, —CSO_3H, —CN, —CHO, —OH, —NH_2.

36. Name a functional group which never becomes a suffix.

37. [i] What is wrong in tricyclo [4, 2, 1, 0] octane? [ii] Draw the structure of tricyclo [4:3:0:$0^{2,4}$] nonane. [iii] How many digits will be present in [ ] in a 'tetracyclo' compound? [iv] Draw the structure of bicyclo [3:3:0] 2-oxooctane. [v] Draw the sructure of spiro [2, 3] hex-1-ene. [vi] between azabenzene and diazabenzene which one requires a locant? [vii] What is the structure of 3-aminopropanoic acid lactam? Is it proper to name this compound as 2-azacyclobutanone?

38. Draw all the isomers of an alcohol $C_5H_{12}O$ and give the IUPAC names of all of them and indicate which compounds will be optically active.

39. Why D-erythro-L-idooctose is considered as the 'relative' IUPAC nomenclature? Is there any difficulty in writing the structure from this name? What should be 'absolute' nomenclature of it?

40. Show that 2-methylbutan-3-ol and 3,3-dimethylbutan-2-ol are correct nomenclatures.

41. Why butyne, aminopropane, dichloroethane, aminobutenyne will exhibit isomerism?

42. Oil of niobe and oil of garlic are what substances?

43. What is power alcohol?

44. What is the difference between a thermosetting and a thermoplastic polymer?

45. What is the difference between CNG and LPG?

46. Name the following:

[i] $C_2H_5^+$, [ii] $CH_3—CH_2—CH_2^-$, [iii] KOBu^t, [iv] $CH_3—CH_2—CH_2—CO_2K$, and

[v] EtO—$(CH_2)_2$—O—$(CH_2)_2$.
                    |
HOOC—CH_2—O

47. Identify the stereochemical notations of the following compounds:

[i] 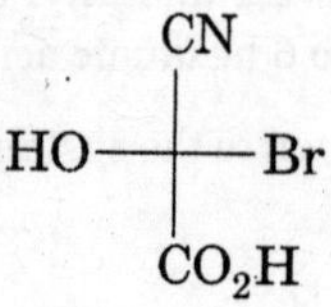

R or *S*

[ii] 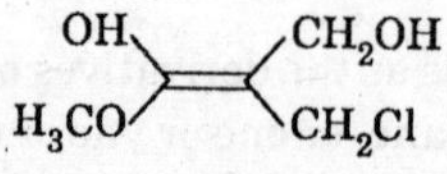

E or *Z*

[iii] 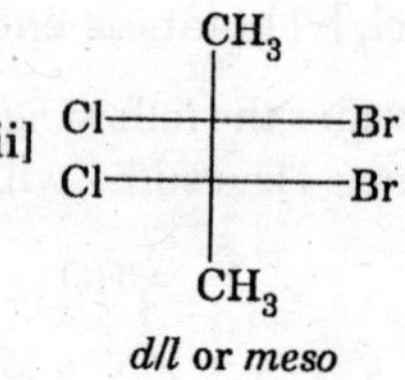

d/l or *meso*

[iv] 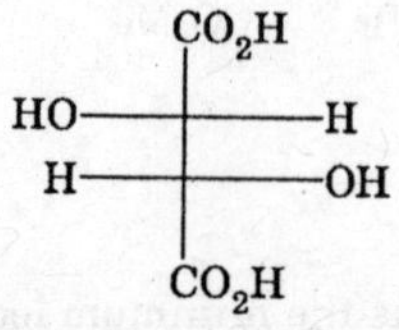

D(+) or L(–)-Tartaric acid

[v] 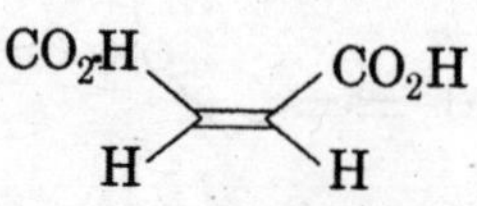

re, *si* or *re*, *re face*

48. Draw the classical structures of (i) *t-a-t*-perhydrophenanthrene, (ii) D-(+)-Glucose, (iii) 2*R*, 3*S*, 4*S*, 5-Tetrahydroxyphentanal.

49. Give one example of each one of the following classes of compounds.
(i) Catenane, (ii) Knot, (iii) Ansa, (iv) Crown ether, (v) Cyclophane, (vi) Allene.

50. Draw the Fischer and Newman projections of D-(–)-erythrose.

51. (a) The innumerable three dimensional pictures of a molecule derived by rotation of a single bond (σ-bond) are called *conformers* of which only a few are capable of finite existence as shown below. The conformers are restricted to only sp^3 carbons. The changes like (α, β) to (β, α) both *trans* and (α, α) to (β, β) both *cis* are the *conformational* changes while (α, α) to (α, β) is a *configurational* (i.e. *cis* to *trans*) change.
The stable conformations are shown below taking X and Y as reference ligands in the Newman projections.

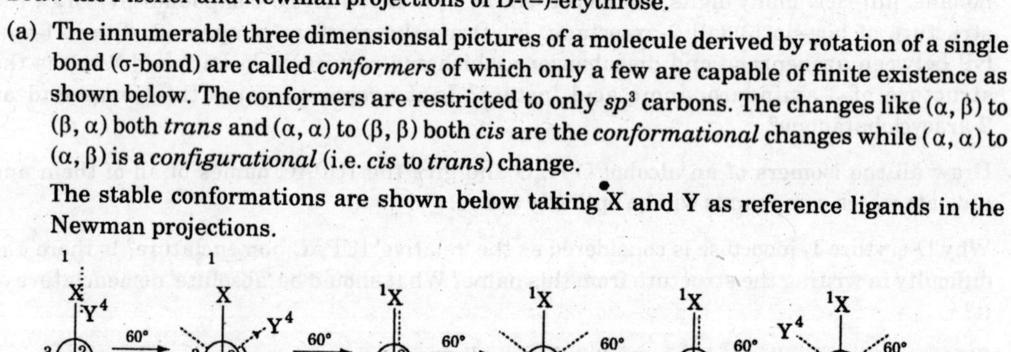

[A] Fully eclipsed — Gauche (pr. gosh) — Partially eclipsed — Staggered — Partially eclipsed — Gauche

Now draw the staggered and gauche *conformations* of D-(–)- *erythrose* and D-(–)-threose.

(b) "The rigid structure of a stereoisomer changeable by exchange of positions of any two ligands is called *configuration*. Thus, *cis/trans*, *E/Z*, *R/S*, etc are *configurational* notations. The changeable (α, β or β, α) to (α, α or β, β) is a *configurational* change as the former is *trans* and the latter is *cis*- isomers. The *conformational* change does not deliver a new isomer."

Criticise this statement with apposite examples.

INDEX

B

D

E

F

G

H

I

T

U

V

X

Y

Z

Some exciting gifts are up for grabs with this book!!

NOMENCLATURE OF THE ORGANIC COMPOUNDS FOR THE BEGINNERS

FREE Gifts Rs. 1000/-

Tanya Das

Amalendu Banerjee

DEAR READERS

By now you must have gone through this book in the entirety. So we invite your assessment on this title with absolute impartiality and candid observations. After all, we publishers blindly depend on your benchmarking. Friends, please fill up the open spaces below with CAPITAL LETTERS. You may add some extra sheets of paper if the space provided is not sufficient for you.

i. What are the important topics of your syllabus which are not adequately covered in this book?

ii. In which chapters and on what topics the treatment of the subject-matter by the author is not satisfactory and why?

iii. Have you come across the typos, factual inaccuracies, inconsistencies, and other forms of mistakes in the book? If yes, then please specify the mistakes and page numbers.

iv. Name top three books on the same subject, which, in your consideration, are better than this book. Could you provide some reasons?

Your 1st Preference: ....

Your 2nd Preference: ....

Your 3rd Preference: ....

v. Any suggestion to improve the quality of illustrations, diagrams, etc.?

vi. How did you like the cover design? Any suggestion on colours?

vii. Suggestions for further improvement of production-standard:

viii. Who recommended you this book:

☐ Teacher ☐ Friend ☐ Parents ☐ Bookseller ☐ Librarian

ix. Name of the person who has recommended:

Designation: ....

Address: ....

Phone No.: ....

e-mail: ....

x. Name and address of the bookseller from whom you have purchased this book:

....

....

....

xi. Name and address of the library where you would like to find the book:

....

....

xii. Name and address of your institution (Please mention the University or Board):

....

....

xiii. Your name, e-mail, and complete postal address:

....

....

....

xiv. A few other titles published by us that you would like to read:

....

....

The best assessment will be rewarded half-yearly. The award will be in the form of our publications, as decided by the Editorial and Marketing Operations Division, amounting to Rs. 1000/- (Rupees one thousand) only.

....

Please mail the filled up coupon at your earliest to :

Editorial & Marketing Operations Division

NEW CENTRAL BOOK AGENCY (P) Ltd

8/1 Chintamoni Das Lane, Kolkata 700 009

West Bengal, INDIA

Phone + Fax: 033 + 2241 8411, 2241 8573, 2241 9094

e-mail: ncbapvtltd@eth.net, ncba@cal2.vsnl.net.in

Website: www.centralgroup.com